PUNISHMENT AND RESPON

PUNISHMENT
and
RESPONSIBILITY

Essays in the Philosophy of Law

BY

H. L. A. HART

PRINCIPAL OF BRASENOSE COLLEGE
OXFORD

CLARENDON PRESS · OXFORD

Oxford University Press, Walton Street, Oxford OX2 6DP

OXFORD LONDON GLASGOW

NEW YORK TORONTO MELBOURNE WELLINGTON

IBADAN NAIROBI DAR ES SALAAM LUSAKA CAPE TOWN

KUALA LUMPUR SINGAPORE JAKARTA HONG KONG TOKYO

DELHI BOMBAY CALCUTTA MADRAS KARACHI

ISBN 0 19 825181 5

First published 1968
Reprinted (with revisions) 1970, 1973, 1978

Printed in Great Britain
at the University Press, Oxford
by Vivian Ridler
Printer to the University

PREFACE

These nine essays, written during the last ten years, are arranged here, not in chronological order, but with some attention to similarity of subject-matter. Chapters I, II and III are each in part concerned with the rationale of the doctrine of *mens rea*, though my main reason for reprinting the essay forming Chapter III is its relevance, in conjunction with the later legal and statistical material added in the Notes, to the discussion of capital punishment which will shortly again be an issue in this country, since the period of abolition secured by the Murder Act 1965 runs out in 1970. Chapters IV, V and VI are each mainly concerned with the analysis of a specific condition of criminal responsibility ('acts of will', intention, and negligence). Chapters VII and VIII confront the claim that the criminal law could and should dispense with mental conditions of responsibility. In Chapter IX, I attempt to distinguish and relate the bewilderingly many meanings of 'responsibility' and 'retribution'.

I have not reprinted here, in spite of some requests, my earliest venture into this field: 'The Ascription of Responsibility and Rights', published in the *Proceedings of the Aristotelian Society* (1948–9). My reason for excluding it is simply that its main contentions no longer seem to me defensible, and that the main criticisms of it made in recent years are justified.[1]

I am deeply indebted to Mr. A. J. Baxter who identified and corrected more inaccurate quotations and references, and more infelicities of grammar, punctuation and style, than I care to contemplate. I am responsible for those that remain. I am also grateful to Miss Joan Watson for her assistance in the preparation of the Notes.

University College
Oxford
1967

H.L.A.H.

[1] See, for example, P. T. Geach, 'Ascriptivism', *Philosophical Review* (1960), p. 221, and G. Pitcher, 'Hart on Action and Responsibility', ibid., p. 266.

NOTE TO SECOND IMPRESSION

In December 1969 both houses of Parliament passed the resolutions required to prevent the expiry of the Murder Act, 1965. I have taken the opportunity presented by a second impression of these essays to add an Appendix on the main statistical arguments used in the Parliamentary debates, and I have brought the statistics for England and Wales and the U.S.A. up to 1968. I have also corrected a number of minor errors, mainly in the Notes, and accommodated the more important changes in the law since the first impression.

H. L. A. H.

Oxford, January 1970

ACKNOWLEDGEMENTS

With the exception of part of Chapter IX, all the essays in this book have been previously published and are here reprinted with minor alterations. The Notes on pp. 238–67, with the exception of the note on Holmes's theory of objective liability on pp. 242–4 and the note on responsibility on pp. 264–5, have not been previously published.

Chapter I, 'Prolegomenon to the Principles of Punishment', was delivered as the Presidential Address to the Aristotelian Society on 19 October 1959, and is reprinted by permission of the Editor. Chapter II, 'Legal Responsibility and Excuses', from the proceedings of the first annual New York Institute of Philosophy, 9 and 10 February 1957, was published in *Determinism and Freedom* (1958) edited by Sidney Hook, and is reprinted by permission of the publishers, New York University Press. Chapter III, 'Murder and the Principles of Punishment: England and the United States', is reprinted by special permission of the *Northwestern University Law Review*, copyright © 1958 by Northwestern University School of Law, Vol. 52 No. 4. Chapter IV, 'Acts of Will and Responsibility', is reprinted from *The Jubilee Lectures of the Faculty of Law, University of Sheffield* (1960), edited by O. R. Marshall, by permission of the publishers Stevens & Sons and of the Faculty of Law, University of Sheffield. Chapter V, 'Intention and Punishment', is reprinted from the *Oxford Review* No 4, February 1967, by permission of the Editor. Chapter VI, 'Negligence, *Mens Rea* and Criminal Responsibility', is reprinted from *Oxford Essays in Jurisprudence* (1961), edited by A. G. Guest, by permission of the Editor and of the publishers, Oxford University Press. Chapter VII, 'Punishment and the Elimination of Responsibility' was delivered as the Hobhouse Memorial Trust Lecture on 16 May 1961 at King's College, London, and is reprinted by permission of the Athlone Press, who published it in 1962, and of the London School of Economics.

Chapter VIII, 'Changing Conceptions of Responsibility', one of two Lionel Cohen Lectures delivered at the Hebrew University, Jerusalem in 1964, was published in *The Morality of the Criminal Law* (1965) and is reprinted by permission of the Magnes Press and Oxford University Press. The first part of Chapter IX, on Responsibility was published in the *Law Quarterly Review* (1967), Vol 83, and is reprinted by permission of the Editor. The note on pp. 242–4, on Holmes's theory of objective liability, is reprinted from *The New York Review of Books*, copyright © 1963 The New York Review, by permission of the Editor.

CONTENTS

I

PROLEGOMENON TO THE PRINCIPLES OF PUNISHMENT

1. INTRODUCTORY

THE main object of this paper is to provide a framework for the discussion of the mounting perplexities which now surround the institution of criminal punishment, and to show that any morally tolerable account of this institution must exhibit it as a compromise between distinct and partly conflicting principles.

General interest in the topic of punishment has never been greater than it is at present and I doubt if the public discussion of it has ever been more confused. The interest and the confusion are both in part due to relatively modern scepticism about two elements which have figured as essential parts of the traditionally opposed 'theories' of punishment. On the one hand, the old Benthamite confidence in fear of the penalties threatened by the law as a powerful deterrent, has waned with the growing realization that the part played by calculation of any sort in anti-social behaviour has been exaggerated. On the other hand a cloud of doubt has settled over the keystone of 'retributive' theory. Its advocates can no longer speak with the old confidence that statements of the form 'This man who has broken the law could have kept it' had a univocal or agreed meaning; or where scepticism does not attach to the *meaning* of this form of statement, it has shaken the confidence that we are generally able to distinguish the cases where a statement of this form is true from those where it is not.[1]

Yet quite apart from the uncertainty engendered by these

[1] See Barbara Wootton *Social Science and Social Pathology* (1959) for a comprehensive modern statement of these doubts.

fundamental doubts, which seem to call in question the accounts given of the efficacy, and the morality of punishment by all the old competing theories, the public utterances of those who conceive themselves to be expounding, as plain men for other plain men, orthodox or common-sense principles (untouched by modern psychological doubts) are uneasy. Their words often sound as if the authors had not fully grasped their meaning or did not intend the words to be taken quite literally. A glance at the parliamentary debates or the *Report of the Royal Commission on Capital Punishment*[2] shows that many are now troubled by the suspicion that the view that there is just one supreme value or objective (e.g. Deterrence, Retribution or Reform) in terms of which *all* questions about the justification of punishment are to be answered, is somehow wrong; yet, from what is said on such occasions no clear account of what the different values or objectives are, or how they fit together in the justification of punishment, can be extracted.[3]

No one expects judges or statesmen occupied in the business of sending people to the gallows or prison, or in making (or unmaking) laws which enable this to be done, to have much time for philosophical discussion of the principles which make it morally tolerable to do these things. A judicial bench is not and should not be a professorial chair. Yet what is said in public debates about punishment by those specially concerned with it as judges or legislators is important. Few are likely to be more circumspect, and if what they say seems, as it often does, unclear, one-sided and easily refutable by pointing to some aspect of things which they have overlooked, it is likely

[2] (1953) Cmd. 8932.

[3] In the Lords' debate in July 1956 the Lord Chancellor agreed with Lord Denning that 'the ultimate justification of any punishment is not that it is a deterrent but that it is the emphatic denunciation by the community of a crime' yet also said that 'the real crux' of the question at issue is whether capital punishment is a uniquely effective deterrent. See 198 *H. L. Deb* (5th July) 576, 577, 596 (1956). In his article, 'An Approach to the Problems of Punishment', *Philosophy* (1958), Mr. S. I. Benn rightly observes of Lord Denning's view that denunciation does not imply the deliberate imposition of suffering which is the feature needing justification (p. 328, n.1).

that in our inherited ways of talking or thinking about punishment there is some persistent drive towards an over-simplification of multiple issues which require separate consideration. To counter this drive what is most needed is *not* the simple admission that instead of a single value or aim (Deterrence, Retribution, Reform or any other) a plurality of different values and aims should be given as a conjunctive answer to some *single* question concerning the justification of punishment. What is needed is the realization that different principles (each of which may in a sense be called a 'justification') are relevant at different points in any morally acceptable account of punishment. What we should look for are answers to a number of different questions such as: What justifies the general practice of punishment? To whom may punishment be applied? How severely may we punish? In dealing with these and other questions concerning punishment we should bear in mind that in this, as in most other social institutions, the pursuit of one aim may be qualified by or provide an opportunity, not to be missed, for the pursuit of others. Till we have developed this sense of the complexity of punishment (and this prolegomenon aims only to do this) we shall be in no fit state to assess the extent to which the whole institution has been eroded by, or needs to be adapted to, new beliefs about the human mind.

2. JUSTIFYING AIMS AND PRINCIPLES OF DISTRIBUTION

There is, I think, an analogy worth considering between the concept of punishment and that of property. In both cases we have to do with a social institution of which the centrally important form is a structure of *legal* rules, even if it would be dogmatic to deny the names of punishment or property to the similar though more rudimentary rule-regulated practices within groups such as a family, or a school, or in customary societies whose customs may lack some of the standard or salient features of law (e.g. legislation, organized sanctions, courts). In both cases we are confronted by a complex institution presenting different inter-related features calling for separate explanation; or, if the morality of the institution is challenged, for separate justification. In both cases failure to distinguish separate questions or attempting to answer them all by reference to a single principle ends in confusion. Thus in the case of property we should distinguish between the question of the *definition* of property, the question why and in what circumstance it is a *good* institution to maintain, and the questions in what ways individuals may become *entitled* to acquire property and *how much* they should be allowed to acquire. These we may call questions of *Definition*, *General Justifying Aim*, and *Distribution* with the last subdivided into questions of *Title* and *Amount*. It is salutary to take some classical exposition of the idea of property, say Locke's chapter 'Of Property' in the *Second Treatise*,[4] and to observe how much darkness is spread by the use of a single notion (in this case 'the labour of (a man's) body and the work of his hands') to answer all these different questions which press upon us when we reflect on the institution of property. In the case of punishment the beginning of wisdom (though by no means its end) is to distinguish similar questions and confront them separately.

(*a*) *Definition*

Here I shall simply draw upon the recent admirable work scattered through English philosophical[5] journals and add to it only an admonition of my own against the abuse of definition in the philosophical discussion of punishment. So with Mr. Benn and Professor Flew I shall define the standard or central case of 'punishment' in terms of five elements:

(i) It must involve pain or other consequences normally considered unpleasant.

[4] Chapter V.

[5] K. Baier, 'Is Punishment Retributive?', *Analysis* (1955), p. 25. A. Flew, 'The Justification of Punishment', *Philosophy* (1954), p. 291. S. I. Benn, op. cit., pp. 325–6.

(ii) It must be for an offence against legal rules.
(iii) It must be of an actual or supposed offender for his offence.
(iv) It must be intentionally administered by human beings other than the offender.
(v) It must be imposed and administered by an authority constituted by a legal system against which the offence is committed.

In calling this the standard or central case of punishment I shall relegate to the position of sub-standard or secondary cases the following among many other possibilities:

(a) Punishments for breaches of legal rules imposed or administered otherwise than by officials (decentralised sanctions).
(b) Punishments for breaches of non-legal rules or orders (punishments in a family or school).
(c) Vicarious or collective punishment of some member of a social group for actions done by others without the former's authorization, encouragement, control or permission.
(d) Punishment of persons (otherwise than under (*c*)) who neither are in fact nor supposed to be offenders.

The chief importance of listing these sub-standard cases is to prevent the use of what I shall call the 'definitional stop' in discussions of punishment. This is an abuse of definition especially tempting when use is made of conditions (ii) and (iii) of the standard case in arguing against the utilitarian claim that the practice of punishment is justified by the beneficial consequences resulting from the observance of the laws which it secures. Here the stock 'retributive' argument[6] is: If *this* is the justification of punishment, why not apply it, when

[6] A. C. Ewing, *The Morality of Punishment*, D. J. B. Hawkins, *Punishment and Moral Responsibility* (The King's Good Servant, p. 92), J. D. Mabbott, 'Punishment', *Mind* (1939), p. 152.

it pays to do so, to those innocent of any crime, chosen at random, or to the wife and children of the offender? And here the wrong reply is: *That*, by definition, would not be 'punishment' and it is the justification of punishment which is in issue.[7] Not only will this definitional stop fail to satisfy the advocate of 'Retribution', it would prevent us from investigating the very thing which modern scepticism most calls in question: namely the rational and moral status of our preference for a system of punishment under which measures painful to individuals are to be taken against them only when they have committed an offence. Why do we prefer this to other forms of social hygiene which we might employ to prevent anti-social behaviour and which we do employ in special circumstances, sometimes with reluctance? No account of punishment can afford to dismiss this question with a definition.

(*b*) *The nature of an offence*

Before we reach any question of justification we must identify a preliminary question to which the answer is so simple that the question may not appear worth asking; yet it is clear that some curious 'theories' of punishment gain their only plausibility from ignoring it, and others from confusing it with other questions. This question is: Why are certain kinds of action forbidden by law and so made crimes or offences? The answer is: To announce to society that these actions are not to be done and to secure that fewer of them are done. These are the common immediate aims of making any conduct a criminal offence and until we have laws made with these primary aims we shall lack the notion of a 'crime' and so of a 'criminal'. Without recourse to the simple idea that the criminal law sets up, in its rules, standards of behaviour to encourage certain types of conduct and discourage others we cannot distinguish a punishment in the form of a fine from a tax on a course of

[7] Mr. Benn seemed to succumb at times to the temptation to give 'The short answer to the critics of utilitarian theories of punishment—that they are theories of *punishment* not of any sort of technique involving suffering' (op. cit., p. 332). He has since told me that he does not now rely on the definitional stop.

conduct.[8] This indeed is one grave objection to those theories of law which in the interests of simplicity or uniformity obscure the distinction between primary laws setting standards for behaviour and secondary laws specifying what officials must or may do when they are broken. Such theories insist that all legal rules are 'really' directions to officials to exact 'sanctions' under certain conditions, e.g. if people kill.[9] Yet only if we keep alive the distinction (which such theories thus obscure) between the primary objective of the law in encouraging or discouraging certain kinds of behaviour, and its merely ancillary sanction or remedial steps, can we give sense to the notion of a crime or offence.

It is important however to stress the fact that in thus identifying the immediate aims of the criminal law we have not reached the stage of justification. There are indeed many forms of undesirable behaviour which it would be foolish (because ineffective or too costly) to attempt to inhibit by use of the law and some of these may be better left to educators, trades unions, churches, marriage guidance councils or other non-legal agencies. Conversely there are some forms of conduct which we believe cannot be effectively inhibited without use of the law. But it is only too plain that in fact the law may make activities criminal which it is morally important to promote and the suppression of these may be quite unjustifiable. Yet confusion between the simple immediate aim of any criminal legislation and the justification of punishment seems to be the most charitable explanation of the claim that punishment is *justified* as an 'emphatic denunciation by the community of a crime'. Lord Denning's dictum that this is the ultimate justification of punishment[10] can be saved from Mr. Benn's

[8] This generally clear distinction may be blurred. Taxes may be imposed to discourage the activities taxed though the law does not announce this as it does when it makes them criminal. Conversely fines payable for some criminal offences because of a depreciation of currency become so small that they are cheerfully paid and offences are frequent. They are then felt to be mere taxes because the sense is lost that the rule is meant to be taken seriously as a standard of behaviour.

[9] cf. Kelsen, *General Theory of Law and State* (1945), pp. 30–33, 33–34, 143–4. 'Law is the primary norm, which stipulates the sanction. . . .' (ibid. 61).

[10] In evidence to the Royal Commission on Capital Punishment, Cmd. 8932. para. 53 (1953). *Supra,* p. 2, n. 3.

criticism, noted above, only if it is treated as a blurred statement of the truth that the aim not of punishment, but of criminal legislation is indeed to denounce certain types of conduct as something not to be practised. Conversely the immediate aim of criminal legislation cannot be any of the things which are usually mentioned as justifying punishment: for until it is settled what conduct is to be legally denounced and discouraged we have not settled from what we are to *deter* people, or who are to be considered *criminals* from whom we are to exact *retribution*, or on whom we are to wreak *vengeance*, or whom we are to *reform*.

Even those who look upon human law as a mere instrument for enforcing 'morality as such' (itself conceived as the law of God or Nature) and who at the stage of justifying punishment wish to appeal not to socially beneficial consequences but simply to the intrinsic value of inflicting suffering on wrongdoers who have disturbed by their offence the moral order, would not deny that the aim of criminal legislation is to set up types of behaviour (in this case conformity with a pre-existing moral law) as legal standards of behaviour and to secure conformity with them. No doubt in all communities certain moral offences, e.g. killing, will always be selected for suppression as crimes and it is conceivable that this may be done not to protect human beings from being killed but to save the potential murderer from sin; but it would be paradoxical to look upon the law as designed not to discourage murder at all (even conceived as sin rather than harm) but simply to extract the penalty from the murderer.

(*c*) *General Justifying Aim*

I shall not here criticize the intelligibility or consistency or adequacy of those theories that are united in denying that the practice of a system of punishment is justified by its beneficial consequences and claim instead that the main justification of the practice lies in the fact that when breach of the law involves moral guilt the application to the offender of the pain of punishment is itself a thing of value. A great variety of claims of this character, designating 'Retribution' or 'Expiation' or 'Reprobation' as the justifying aim, fall in spite of differences under this rough general description. Though in fact I agree with Mr. Benn[11] in thinking that these all either avoid the question of justification altogether or are in spite of their protestations disguised forms of Utilitarianism, I shall assume that Retribution, defined simply as the application of the pains of punishment to an offender who is morally guilty, may figure among the conceivable justifying aims of a system of punishment. Here I shall merely insist that it is one thing to use the word Retribution *at this point* in an account of the principle of punishment in order to designate the General Justifying Aim of the system, and quite another to use it to secure that to the question 'To whom may punishment be applied?' (the question of Distribution), the answer given is 'Only to an offender for an offence'. Failure to distinguish Retribution as a General Justifying Aim from retribution as the simple insistence that only those who have broken the law —and voluntarily broken it—may be punished, may be traced in many writers: even perhaps in Mr. J. D. Mabbott's[12] otherwise most illuminating essay. We shall distinguish the latter from Retribution in General Aim as 'retribution in Distribution'. Much confusing shadow-fighting between utilitarians and their opponents may be avoided if it is recognized that it is perfectly consistent to assert *both* that the General Justifying Aim of the practice of punishment is its beneficial consequences *and* that the pursuit of this General Aim should be qualified or restricted out of deference to principles of Distribution which require that punishment should be only of an offender for an offence. Conversely it does not in the least follow from the admission of the latter principle of retribution in Distribution that the General Justifying Aim of punishment is Retribution though of course Retribution in General Aim entails retribution in Distribution.

We shall consider later the principles of justice lying at the

[11] Op. cit., pp. 326–35.

[12] Op. cit. *supra* p. 5, n. 6. It is not always quite clear what he considers a 'retributive' theory to be.

root of retribution in Distribution. Meanwhile it is worth observing that both the old fashioned Retributionist (in General Aim) and the most modern sceptic often make the same (and, I think, wholly mistaken) assumption that sense can only be made of the restrictive principle that punishment be applied only to an offender for an offence if the General Justifying Aim of the practice of punishment is Retribution. The sceptic consequently imputes to all systems of punishment (when they are restricted by the principle of retribution in Distribution) all the irrationality he finds in the idea of Retribution as a General Justifying Aim; conversely the advocates of the latter think the admission of retribution in Distribution is a refutation of the utilitarian claim that the social consequences of punishment are its Justifying Aim.

The most general lesson to be learnt from this extends beyond the topic of punishment. It is, that in relation to any social institution, after stating what general aim or value its maintenance fosters we should enquire whether there are any and if so what principles limiting the unqualified pursuit of that aim or value. Just because the pursuit of any single social aim always has its restrictive qualifier, our main social institutions always possess a plurality of features which can only be understood as a compromise between partly discrepant principles. This is true even of relatively minor legal institutions like that of a contract. In general this is designed to enable individuals to give effect to their wishes to create structures of legal rights and duties, and so to change, in certain ways, their legal position. Yet at the same time there is need to protect those who, in good faith, understand a verbal offer made to them to mean what it would ordinarily mean, accept it, and then act on the footing that a valid contract has been concluded. As against them, it would be unfair to allow the other party to say that the words he used in his verbal offer or the interpretation put on them did not express his real wishes or intention. Hence principles of 'estoppel' or doctrines of the 'objective sense' of a contract are introduced to prevent this and to qualify the principle that the law enforces contracts in order to give effect to the joint wishes of the contracting parties.

(*d*) *Distribution*

This as in the case of property has two aspects (i) Liability (Who may be punished?) and (ii) Amount. In this section I shall chiefly be concerned with the first of these.[13]

From the foregoing discussions two things emerge. First, though we may be clear as to what value the practice of punishment is to promote, we have still to answer as a question of Distribution 'Who may be punished?' Secondly, if in answer to this question we say 'only an offender for an offence' this admission of retribution in Distribution is not a principle from which anything follows as to the severity or amount of punishment; in particular it neither licenses nor requires, as Retribution in General Aim does, more severe punishments than deterrence or other utilitarian criteria would require.

The root question to be considered is, however, why we attach the moral importance which we do to retribution in Distribution. Here I shall consider the efforts made to show that restriction of punishment to offenders is a simple consequence of whatever principles (Retributive or Utilitarian) constitute the Justifying Aim of punishment.

The standard example used by philosophers to bring out the importance of retribution in Distribution is that of a wholly innocent person who has not even unintentionally done anything which the law punishes if done intentionally. It is supposed that in order to avert some social catastrophe officials of the system fabricate evidence on which he is charged, tried, convicted and sent to prison or death. Or it is supposed that without resort to any fraud more persons may be deterred from crime if wives and children of offenders were punished vicariously for their crimes. In some forms this kind of thing may be ruled out by a consistent sufficiently comprehensive utilitarianism.[14] Certainly expedients involving fraud or faked

[13] Amount is considered below in Section III (in connexion with Mitigation) and Section V.

[14] See J. Rawls, 'Two Concepts of Rules', *Philosophical Review* (1955), pp. 4–13.

charges might be very difficult to justify on utilitarian grounds. We can of course imagine that a negro might be sent to prison or executed on a false charge of rape in order to avoid widespread lynching of many others; but a *system* which openly empowered authorities to do this kind of thing, even if it succeeded in averting specific evils like lynching, would awaken such apprehension and insecurity that any gain from the exercise of these powers would by any utilitarian calculation be offset by the misery caused by their existence. But official resort to this kind of fraud on a particular occasion in breach of the rules and the subsequent indemnification of the officials responsible might save many lives and so be thought to yield a clear surplus of value. Certainly vicarious punishment of an offender's family might do so and legal systems have occasionally though exceptionally resorted to this. An example of it is the Roman *Lex Quisquis* providing for the punishment of the children of those guilty of *majestas*.[15] In extreme cases many might still think it right to resort to these expedients but we should do so with the sense of sacrificing an important principle. We should be conscious of choosing the lesser of two evils, and this would be inexplicable if the principle sacrificed to utility were itself only a requirement of utility.

Similarly the moral importance of the restriction of punishment to the offender cannot be explained as merely a consequence of the principle that the General Justifying Aim is Retribution for immorality involved in breaking the law. Retribution in the Distribution of punishment has a value quite independent of Retribution as Justifying Aim. This is shown by the fact that we attach importance to the restrictive principle that only offenders may be punished, even where breach of this law might not be thought immoral. Indeed even where the laws themselves are hideously immoral as in Nazi Germany, e.g. forbidding activities (helping the sick or destitute of some racial group) which might be thought morally obligatory, the absence of the principle restricting punishment to the offender would be a further *special* iniquity; whereas admission of this principle would represent some

[15] Constitution of emperors Arcadius and Honorius (A.D. 397).

residual respect for justice shown in the administration of morally bad laws.

3. JUSTIFICATION, EXCUSE AND MITIGATION

What is morally at stake in the restrictive principle of Distribution cannot, however, be made clear by these isolated examples of its violation by faked charges or vicarious punishment. To make it clear we must allot to their place the appeals to matters of Justification, Excuse and Mitigation made in answer to the claim that someone should be punished. The first of these depends on the General Justifying Aim; the last two are different aspects of the principles of Distribution of punishment.

(a) *Justification and Excuse*

English lawyers once distinguished between 'excusable' homicide (e.g. accidental non-negligent killing) and 'justifiable homicide (e.g. killing in self-defence or in the arrest of a felon) and different legal consequences once attached to these two forms of homicide. To the modern lawyer this distinction has no longer any legal importance: he would simply consider both kinds of homicide to be cases where some element, negative or positive, required in the full definition of criminal homicide (murder or manslaughter) was lacking. But the distinction between these two different ways in which actions may fail to constitute a criminal offence is still of great moral importance. Killing in self-defence is an exception to a general rule making killing punishable; it is admitted because the policy or aims which in general justify the punishment of killing (e.g. protection of human life) do not include cases such as this. In the case of 'justification' what is done is regarded as something which the law does not condemn, or even welcomes.[16] But where killing (e.g. accidental) is excused,

[16] In 1811 Mr. Purcell of Co. Cork, a septuagenarian, was knighted for killing four burglars with a carving knife. Kenny, *Outlines of Criminal Law*, 5th edn., p. 103, n. 3.

criminal responsibility is excluded on a different footing. What has been done is something which is deplored, but the psychological state of the agent when he did it exemplified one or more of a variety of conditions which are held to rule out the public condemnation and punishment of individuals. This is a requirement of fairness or of justice to individuals independent of whatever the General Aim of punishment is, and remains a value whether the laws are good, morally indifferent or iniquitous.

The most prominent of these excusing conditions are those forms of lack of knowledge which make action unintentional: lack of muscular control which makes it involuntary, subjection to gross forms of coercion by threats, and types of mental abnormality, which are believed to render the agent incapable of choice or of carrying out what he has chosen to do. Not all these excusing conditions are admitted by all legal systems for all offenders. Nearly all penal systems, as we shall see, make some compromise at this point with other principles; but most of them are admitted to some considerable extent in the case of the most serious crimes. Actions done under these excusing conditions are in the misleading terminology of Anglo-American law done without *mens rea*,[17] and most people would say of them that they were not 'voluntary' or 'not wholly voluntary'.

(*b*) *Mitigation*

Justification and Excuse though different from each other are alike in that if either is made out then conviction and punishment are excluded. In this they differ from the idea of Mitigation which presupposes that someone is convicted and liable to be punished and the question of the severity of his punishment is to be decided. It is therefore relevant to that aspect of Distribution which we have termed amount. Certainly the severity of punishment is in part determined by the General Justifying Aim. A utilitarian will for example exclude in principle punishments the infliction of which is

[17] Misleading because it suggests moral guilt is a necessary condition of criminal responsibility.

held to cause more suffering than the offence unchecked, and will hold that if one kind of crime causes greater suffering than another then a greater penalty may be used, if necessary, to repress it. He will also exclude degrees of severity which are useless in the sense that they do no more to secure or maintain a higher level of law-observance or any other valued result than less severe penalties. But in addition to restrictions on the severity of punishment which follow from the aim of punishing, special limitations are imported by the idea of Mitigation. These, like the principle of Distribution restricting liability to punishment to offenders, have a status which is independent of the general Aim. The special features of Mitigation are that a good reason for administering a less severe penalty is made out if the situation or mental state of the convicted criminal is such that he was exposed to an unusual or specially great temptation, or his ability to control his actions is thought to have been impaired or weakened otherwise than by his own action, so that conformity to the law which he has broken was a matter of special difficulty for him as compared with normal persons normally placed.

The special features of the idea of Mitigation are however often concealed by the various legal techniques which make it necessary to distinguish between what may be termed 'informal' and 'formal' Mitigation. In the first case the law fixes a maximum penalty and leaves it to the judge to give such weight as he thinks proper, in selecting the punishment to be applied to a particular offender, to (among other considerations) mitigating factors. It is here that the barrister makes his 'plea in mitigation'. Sometimes however legal rules provide that the presence of a mitigating factor shall always remove the offence into a separate category carrying a lower maximum penalty. This is 'formal' mitigation and the most prominent example of it is Provocation which in English law is operative only in relation to homicide. Provocation is not a matter of Justification or Excuse for it does not exclude conviction or punishment; but 'reduces' the charges from murder to manslaughter and the possible maximum penalty from death to life imprisonment. It is worth stressing that not every provision reducing the maximum penalty can be thought of as 'Mitigation': the very peculiar provisions of s. 5 of the Homicide Act 1957 which (*inter alia*) restricted the death penalty to types of murder not including, for example, murder by poisoning, did not in doing this recognize the use of poison as a 'mitigating circumstance'. Only a reduction of penalty made in view of the individual criminal's special difficulties in keeping the law which he has broken is so conceived.

Though the central cases are distinct enough the border lines between Justification, Excuse and Mitigation are not. There are many features of conduct which can be and are thought of in more than one of these ways. Thus, though little is heard of it, duress (coercion by threat of serious harm) is in English law in relation to some crimes an Excuse excluding responsibility. Where it is so treated the conception is that since *B* has committed a crime only because *A* has threatened him with gross violence or other harm, *B's* action is not the outcome of a 'free' or independent choice; *B* is merely an instrument of *A* who has 'made him do it'. Nonetheless *B* is not an instrument in the same sense that he would have been had he been pushed by *A* against a window and broken it: unless he is literally paralysed by fear of the threat, we may believe that *B* could have refused to comply. If he complies we may say '*coactus voluit*' and treat the situation not as one making it intolerable to punish at all, but as one calling for mitigation of the penalty as gross provocation does. On the other hand if the crime which *A* requires *B* to commit is a petty one compared with the serious harm threatened (e.g. death) by *A* there would be no absurdity in treating *A*'s threat as a justification for *B*'s conduct though few legal systems overtly do this. If this line is taken coercion merges into the idea of 'necessity'[18] which appears on the margin of most systems of criminal law as an exculpating factor.

In view of the character of modern sceptical doubts about criminal punishment it is worth observing that, even in English

[18] i.e. when breaking the law is held justified as the lesser of two evils.

law, the relevance of mental disease to criminal punishment is not always as a matter of Excuse though exclusive concentration on the M'Naghten rules relating to the criminal responsibility of the mentally diseased encourages the belief that it is. Even before the Homicide Act 1957 a statute[19] provided that if a mother murdered her child under the age of twelve months while 'the balance of her mind was disturbed' by the processes of birth or lactation she should be guilty only of the felony of infanticide carrying a maximum penalty of life imprisonment. This is to treat mental abnormality as a matter of (formal) Mitigation. Similarly in other cases of homicide the M'Naghten rules relating to certain types of insanity as an Excuse no longer stand alone; now such abnormality of mind as 'substantially impaired [the] mental responsibility'[20] of the accused is a matter of formal mitigation, which like provocation reduces the homicide to the category of manslaughter.

4. THE RATIONALE OF EXCUSES

The admission of excusing conditions is a feature of the Distribution of punishment and it is required by distinct principles of Justice which restrict the extent to which general social aims may be pursued at the cost of individuals. The moral importance attached to these in punishment distinguishes it from other measures which pursue similar aims (e.g. the protection of life, wealth or property) by methods which like punishment are also often unpleasant to the individuals to whom they are applied, e.g. the detention of persons of hostile origin or association in war time, or of the insane, or the compulsory quarantine of persons suffering from infectious disease. To these we resort to avoid damage of a catastrophic character.

Every penal system in the name of some other social value compromises over the admission of excusing conditions and no system goes as far (particularly in cases of mental disease) as

[19] Infanticide Act. 1938. [20] Homicide Act, 1957, s. 2.

many would wish. But it is important (if we are to avoid a superficial but tempting answer to modern scepticism about the meaning or truth of the statement that a criminal could have kept the law which he has broken) to see that our moral preference for a system which does recognize such excuses cannot, any more than our reluctance to engage in the cruder business of false charges or vicarious punishment, be explained by reference to the General Aim which we take to justify the practice of punishment. Here, too, even where the laws appear to us morally iniquitous or where we are uncertain as to their moral character so that breach of law does not entail moral guilt, punishment of those who break the law unintentionally would be an added wrong and refusal to do this some sign of grace.

Retributionists (in General Aim) have not paid much attention to the rationale of this aspect of punishment; they have usually (wrongly) assumed that it has no status except as a corollary of Retribution in General Aim. But Utilitarians have made strenuous, detailed efforts to show that restriction of the use of punishment to those who have voluntarily broken the law is explicable on purely utilitarian lines. Bentham's efforts are the most complete and their failure is an instructive warning to contemporaries.

Bentham's argument was a reply to Blackstone who, in expounding the main excusing conditions recognized in the criminal law of his day[21], claimed that 'all the several pleas and excuses which protect the committer of a forbidden act from punishment which is otherwise annexed thereto may be reduced to this single consideration: the want or defect of *will*' . . . . [and to the principle] 'that to constitute a crime . . . . there must be first, a vitious will' In his Introduction to the Principles of Morals and Legislation[22] under the heading 'Cases unmeet for punishment' Bentham sets out a list of the main excusing conditions similar to Blackstone's; he then undertakes to show that the infliction of punishment on those who have done, while in any of these conditions, what the law forbids 'must be inefficacious: it cannot act so as to prevent

[21] *Commentaries*, Book IV, Chap. II. [22] Chap. XIII esp. para. 9, n. 1.

the mischief'. All the common talk about want or defect of will or lack of a 'vitious' will is, he says, 'nothing to the purpose', except so far as it implies the reason (inefficacy of punishment) which he himself gives for recognising these excuses.

Bentham's argument is in fact a spectacular *non sequitur*. He sets out to prove that to *punish* the mad, the infant child or those who break the law unintentionally or under duress or even under 'necessity' must be inefficacious; but all that he proves (at the most) is the quite different proposition that the *threat* of punishment will be ineffective so far as the class of persons who suffer from these conditions is concerned. Plainly it is possible that though (as Bentham says) the *threat* of punishment could not have operated on them, the actual *infliction* of punishment on those persons, may secure a higher measure of conformity to law on the part of normal persons than is secured by the admission of excusing conditions. If this is so and if Utilitarian principles only were at stake, we should, without any sense that we were sacrificing any principle of value or were choosing the lesser of two evils, drop from the law the restriction on punishment entailed by the admission of excuses: unless, of course, we believed that the terror or insecurity or misery produced by the operation of laws so Draconic was worse than the lower measure of obedience to law secured by the law which admits excuses.

This objection to Bentham's rationale of excuses is not merely a fanciful one. Any increase in the number of conditions required to establish criminal liability increases the opportunity for deceiving courts or juries by the pretence that some condition is not satisfied. When the condition is a psychological factor the chances of such pretence succeeding are considerable. Quite apart from the provision made for mental disease, the cases where an accused person pleads that he killed in his sleep or accidentally or in some temporary abnormal state of unconsciousness show that deception is certainly feasible. From the Utilitarian point of view this may lead to two sorts of "losses". The belief that such deception is feasible may embolden persons who would not otherwise risk punishment to take their chance of deceiving a jury in this way. Secondly, a criminal who actually succeeds in this deception will be left at large, though belonging to the class which the law is concerned to incapacitate. Developments in Anglo-American law since Bentham's day have given more concrete form to this objection to his argument. There are now offences (known as offences of 'strict liability') where it is not necessary for conviction to show that the accused either intentionally did what the law forbids or could have avoided doing it by use of care: selling liquor to an intoxicated person, possessing an altered passport, selling adulterated milk[23] are examples out of a range of 'strict liability' offences where it is no defence that the accused did not offend intentionally, or through negligence, e.g., that he was under some mistake against which he had no opportunity to guard. Two things should be noted about them. First, the common justification of this form of criminal liability is that if proof of intention or lack of care were required guilty persons would escape. Secondly, 'strict liability' is generally viewed with great odium and admitted as an exception to the general rule, with the sense that an important principle has been sacrificed to secure a higher measure of conformity and conviction of offenders. Thus Bentham's argument curiously ignores both the two possibilities which have been realized. First, actual punishment of these who act unintentionally or in some other normally excusing manner may have a utilitarian value in its effects on others; and secondly, when because of this probability, strict liability is admitted and the normal excuses are excluded, this may be done with the sense that some other principle has been overridden.

On this issue modern extended forms of Utilitarianism fare no better than Bentham's whose main criterion here of 'effective' punishment was deterrence of the offender or of others by example. Sometimes the principle that punishment should be restricted to those who have voluntarily broken the law is defended not as a principle which is rational or morally important in itself but as something so engrained in popular conceptions

[23] See Glanville Williams, *Criminal Law,* 2nd edn., Chap. VI, for a discussion of the protest against 'strict responsibility'.

of justice[24] in certain societies, including our own, that not to recognize it would lead to disturbances, or to the nullification of the criminal law since officials or juries might refuse to co-operate in such a system. Hence to punish in these circumstances would either be impracticable or would create more harm than could possibly be offset by any superior deterrent force gained by such a system. On this footing, a system should admit excuses much as, in order to prevent disorder or lynching, concessions might be made to popular demands for more savage punishment than could be defended on other grounds. Two objections confront this wider pragmatic form of Utilitarianism. The first is the factual observation that even if a system of strict liability for all or very serious crime would be unworkable, a system which admits it on its periphery for relatively minor offences is not only workable but an actuality which we have, though many object to it or admit it with reluctance. The second objection is simply that we do not dissociate ourselves from the principle that it is wrong to punish the hopelessly insane or those who act unintentionally, etc., by treating it as something merely embodied in popular *mores* to which concessions must be made sometimes. We condemn legal systems where they disregard this principle; whereas we try to educate people out of their preference for savage penalties even if we might in extreme cases of threatened disorder concede them.

It is therefore impossible to exhibit the principle by which punishment is excluded for those who act under the excusing conditions merely as a corollary of the general Aim—Retributive or Utilitarian—justifying the practice of punishment. Can anything positive be said about this principle except that it is one to which we attach moral importance as a restriction on the pursuit of any aim we have in punishing?

It is clear that like all principles of Justice it is concerned with the adjustment of claims between a multiplicity of persons. It incorporates the idea that each individual person is

[24] Michael and Wechsler, 'A Rationale of the Law of Homicide' (1937), 37 *C.L.R.*, 701, esp. pp. 752–7, and Rawls, op. cit.

to be protected against the claim of the rest for the highest possible measure of security, happiness or welfare which could be got at his expense by condemning him for a breach of the rules and punishing him. For this a moral licence is required in the form of proof that the person punished broke the law by an action which was the outcome of his free choice, and the recognition of excuses is the most we can do to ensure that the terms of the licence are observed. Here perhaps, the elucidation of this restrictive principle should stop. Perhaps we (or I) ought simply to say that it is a requirement of Justice, and Justice simply consists of principles to be observed in adjusting the competing claims of human beings which (i) treat all alike as persons by attaching special significance to human voluntary action and (ii) forbid the use of one human being for the benefit of others except in return for his voluntary actions against them. I confess however to an itch to go further; though what I have to say may not add to these principles of Justice. There are, however, three points which even if they are restatements from different points of view of the principles already stated, may help us to identify what we now think of as values in the practice of punishment and what we may have to reconsider in the light of modern scepticism.

(*a*) We may look upon the principle that punishment must be reserved for voluntary offences from two different points of view. The first is that of the rest of society considered as *harmed* by the offence (either because one of its members has been injured or because the authority of the law essential to its existence has been challenged or both). The principle then appears as one securing that the suffering involved in punishment falls upon those who have voluntarily harmed others: this is valued, not as the Aim of punishment, but as the only fair terms on which the General Aim (protection of society, maintenance of respect for law, etc.) may be pursued.

(*b*) The second point of view is that of society concerned not as harmed by the crime but as *offering* individuals including the criminal the protection of the laws on terms which are fair, because they not only consist of a framework of reciprocal rights and duties, but because within this framework each individual is given a *fair* opportunity to choose between keeping the law required for society's protection or paying the penalty. From the first point of view the actual punishment of a criminal appears not merely as something useful to society (General Aim) but as justly extracted from the criminal who has voluntarily done harm; from the second it appears as a price justly extracted because the criminal had a fair opportunity beforehand to avoid liability to pay.

(*c*) Criminal punishment as an attempt to secure desired behaviour differs from the manipulative techniques of the Brave New World (conditioning, propaganda, etc.) or the simple incapacitation of those with anti-social tendencies, by taking a risk. It defers action till harm has been done; its primary operation consists simply in announcing certain standards of behaviour and attaching penalties for deviation, making it less eligible, and then leaving individuals to choose. This is a method of social control which maximizes individual freedom within the coercive framework of law in a number of different ways, or perhaps, different senses. First, the individual has an option between obeying or paying. The worse the laws are, the more valuable the possibility of exercising this choice becomes in enabling an individual to decide how he shall live. Secondly, this system not only enables individuals to exercise this choice but increases the power of individuals to identify beforehand periods when the law's punishments will not interfere with them and to plan their lives accordingly. This very obvious point is often overshadowed by the other merits of restricting punishment to offences voluntarily committed, but is worth separate attention. Where punishment is not so restricted individuals will be liable to have their plans frustrated by punishments for what they do unintentionally, in ignorance, by accident or mistake. Such a system of strict liability for all offences, if logically possible,[25] would not only vastly increase the number of punishments, but would diminish the individual's power to identify beforehand particular periods during which he will be free from them. This

[25] Some crimes, e.g. demanding money by menaces, cannot (logically) be committed unintentionally.

is so because we can have very little ground for confidence that during a particular period we will not do something unintentionally, accidentally, etc.; whereas from their own knowledge of themselves many can say with justified confidence that for some period ahead they are not likely to engage intentionally in crime and can plan their lives from point to point in confidence that they will be left free during that period. Of course the confidence thus justified, though drawn from knowledge of ourselves, does not amount to certainty. My confidence that I will not during the next twelve months intentionally engage in any crime and will be free from punishment, may turn out to be misplaced; but it is both greater and better justified than my belief that I will not do unintentionally any of the things which our system punishes if done intentionally.

5. REFORM AND THE INDIVIDUALIZATION OF PUNISHMENT

The idea of Mitigation incorporates the conviction that though the amount or severity of punishment is primarily to be determined by reference to the General Aim, yet Justice requires that those who have special difficulties to face in keeping the law which they have broken should be punished less. Principles of Justice however are also widely taken to bear on the amount of punishment in at least two further ways. The first is the somewhat hazy requirement that 'like cases be treated alike'. This is certainly felt to be infringed at least when the ground for different punishment for those guilty of the same crime is neither some personal characteristic of the offender connected with the commission of the crime nor the effect of punishment on him. If a certain offence is specially prevalent at a given time and a judge passes heavier sentences than on previous offenders ('as a warning') some sacrifice of justice to the safety of society is involved though it is often acceptable to many as the lesser of two evils.

The further principle that different kinds of offence of different gravity (however that is assessed) should not be punished with equal severity is one which like other principles of Distribution may qualify the pursuit of our General Aim and is not deducible from it. Long sentences of imprisonment might effectually stamp out car parking offences, yet we think it wrong to employ them; *not* because there is for each crime a penalty 'naturally' fitted to its degree of iniquity (as some Retributionists in General Aim might think); not because we are convinced that the misery caused by such sentences (which might indeed be slight because they would rarely need to be applied) would be greater than that caused by the offences unchecked (as a Utilitarian might argue). The guiding principle is that of a proportion within a system of penalties between those imposed for different offences where these have a distinct place in a commonsense scale of gravity. This scale itself no doubt consists of very broad judgements both of relative moral iniquity and harmfulness of different types of offence: it draws rough distinctions like that between parking offences and homicide, or between 'mercy killing' and murder for gain, but cannot cope with any precise assessment of an individual's wickedness in committing a crime (Who can?) Yet maintenance of proportion of this kind may be important: for where the legal gradation of crimes expressed in the relative severity of penalties diverges sharply from this rough scale, there is a risk of either confusing common morality or flouting it and bringing the law into contempt.

The ideals of Reform and Individualization of punishment (e.g. corrective training, preventive detention) which have been increasingly accepted in English penal practice since 1900 plainly run counter to the second if not to both of these principles of Justice or proportion. Some fear, and others hope, that the further intrusion of these ideals will end with the substitution of 'treatment' by experts for judicial punishment. It is, however, important to see precisely what the relation of Reform to punishment is because its advocates too often misstate it. 'Reform' as an objective is no doubt very vague; it now embraces any strengthening of the offender's disposition and capacity to keep within the law, which is intentionally brought about by human effort otherwise than through fear of punishment. Reforming methods include the inducement of states of repentance, or recognition of moral guilt, or greater awareness of the character and demands of society, the provision of education in a broad sense, vocational training and psychological treatment. Many seeing the futility and indeed harmful character of much traditional punishment speak as if Reform could and should be the General Aim of the whole practice of punishment or the dominant objective of the criminal law:

> 'The *corrective theory* based upon a conception of multiple causation and curative-rehabilitative treatment, should clearly predominate in legislation and in judicial and administrative practices.'[26]

Of course this is a possible ideal but is not an ideal for punishment. Reform can only have a place within a system of punishment as an exploitation of the opportunities presented by the conviction or compulsory detention of offenders. It is not an alternative General Justifying Aim of the practice of punishment but something the pursuit of which within a system of punishment qualifies or displaces altogether recourse to principles of justice or proportion in determining the amount of punishment. This is where both Reform and individualized punishment have run counter to the customary morality of punishment.

There is indeed a paradox in asserting that Reform should 'predominate' in a system of Criminal Law, as if the main purpose of providing punishment for murder was to reform the murderer not to prevent murder; and the paradox is greater where the legal offence is not a serious moral one: e.g. infringing a state monopoly of transport. The objection to assigning to Reform this place in punishment is not merely that punishment entails suffering and Reform does not; but that Reform

[26] Hall and Glueck, *Cases on Criminal Law and its Enforcement* (1951) p. 14.

is essentially a remedial step for which *ex hypothesi* there is an opportunity only at the point where the criminal law has failed in its primary task of securing society from the evil which breach of the law involves. Society is divisible at any moment into two classes (i) those who have actually broken a given law and (ii) those who have not yet broken it but may. To take Reform as the dominant objective would be to forgo the hope of influencing the second and—in relation to the more serious offences—numerically much greater class. We should thus subordinate the prevention of first offences to the prevention of recidivism.

Consideration of what conditions or beliefs would make this appear a reasonable policy brings us to the topic to which this paper is a mere prolegomenon: modern sceptical doubt about the whole institution of punishment. If we believed that nothing was achieved by announcing penalties or by the example of their infliction, either because those who do not commit crimes would not commit them in any event or because the penalties announced or inflicted on others are not among the factors which influence them in keeping the law, then some dramatic change concentrating wholly on actual offenders, would be necessary. Just because at present we do not entirely believe this, we have a dilemma and an uneasy compromise. Penalties which we believe are required as a threat to maintain conformity to law at its maximum may convert the offender to whom they are applied into a hardened enemy of society; while the use of measures of Reform may lower the efficacy and example of punishment on others. At present we compromise on this relatively new aspect of punishment as we do over its main elements. What makes this compromise seem tolerable is the belief that the influence which the threat and example of punishment extracts is often independent of the severity of the punishment, and is due more to the disgrace attached to conviction for crime or to the deprivation of freedom which many reforming measures at present used in any case involve.

II

LEGAL RESPONSIBILITY AND EXCUSES

I

IT is characteristic of our own and all advanced legal systems that the individual's liability to punishment, at any rate for serious crimes carrying severe penalties, is made by law to depend, among other things, on certain mental conditions. These conditions can best be expressed in negative form as *excusing* conditions: the individual is not liable to punishment if at the time of his doing what would otherwise be a punishable act he was unconscious, mistaken about the physical consequences of his bodily movements or the nature or qualities of the thing or persons affected by them, or, in some cases, if he was subjected to threats or other gross forms of coercion or was the victim of certain types of mental disease. This is a list, not meant to be complete, giving broad descriptions of the principal excusing conditions; the exact definition of these and their precise character and scope must be sought in the detailed exposition of our criminal law. If an individual breaks the law when none of the excusing conditions are present he is ordinarily said to have acted of 'his own free will', 'of his his own accord', 'voluntarily'; or it might be said, 'He could have helped doing what he did.' If the determinist[1] has anything to say on this subject, it must be because he makes two

[1] A variety of theories or claims shelter under the label 'determinism'. For many purposes it is necessary to distinguish among them, especially on the question whether the elements in human conduct that are said to be 'determined' are regarded as the product of sufficient conditions, or sets of jointly sufficient conditions, which include the individual's character. I think, however, that the defence I make in this paper of the rationality, morality, and justice of qualifying criminal responsibility by excusing conditions will be compatible with any form of determinism which satisfies the two following sets of requirements.

A. The determinist must not deny (*a*) those *empirical* facts that at present we treat as proper grounds for saying, 'He did what he chose', 'His choice was

claims. The first claim is that it may be true—though we cannot yet show and may never be able to show that it is true—that human conduct (including in that expression not only actions involving the movements of the human body but its psychological elements or components such as decisions, choices, experiences of desire, effort, etc.) are subject to certain types of law, where law is to be understood in the sense of a scientific law. The second claim is that, if human conduct so understood is in fact subject to such laws (though at the present time we do not know it to be so), the distinction we draw between one who acts under excusing conditions and one who acts when none are present becomes unimportant, if not absurd. Consequently, to allow punishment to depend on the presence or absence of excusing conditions, or to think it justified when they are absent but not when they are present, is absurd, meaningless, irrational, or unjust, or immoral, or perhaps all of these.

My principal object in this paper is to draw attention to the analogy between conditions that are treated by criminal law as *excusing* conditions and certain similar conditions that are treated in another branch of the law as *invalidating* certain civil transactions such as wills, gifts, contracts, and marriages. If we consider this analogy, I think we can see that there is a rationale for our insistence on the importance of excusing conditions in criminal law that no form of determinism that I, at any rate, can construct could impugn; and this rationale

effective', 'He got what he chose', 'That was the result of his choice', etc.; (*b*) the fact that when we get what we chose to have, live our lives as we have chosen, and particularly when we obtain by a choice what we have judged to be the lesser of two evils, this is a source of satisfaction; (*c*) the fact that we are often able to predict successfully and on reasonable evidence that our choice will be effective over certain periods in relation to certain matters.

B. The determinist does not assert and could not truly assert that we *already know* the laws that he says may exist or (in some versions) *must* exist. Determinists differ on the question whether or not the laws are sufficiently simple (*a*) for human beings to discover, (*b*) for human beings to use for the prediction of their own and others' conduct. But as long as it is not asserted that we know these laws I do not think this difference of opinion important here. Of course if we knew the laws and could use them for the detailed and exact prediction of our own and others' conduct, *deliberation* and *choice* would become pointless, and perhaps in such circumstances there could not (logically) be 'deliberation' or 'choice.'

seems to me superior at many points to the two main accounts or explanations which in Anglo-American jurisprudence have been put forward as the basis of the recognition of excusing conditions in criminal responsibility.

In this preliminary section, however, I want to explain why I shall not undertake the analysis or elucidation of the meaning of such expressions as 'He did it voluntarily', 'He acted of his own free will', 'He could have helped doing it', 'He could have done otherwise'. I do not, of course, think the analysis of these terms unimportant: indeed I think we owe the progress that has been made, at least in determining what 'the free will problem' is, to the work of philosophers who have pursued this analysis. Perhaps it may be shown that statements of the form 'He did it of his own free will' or 'He could have done otherwise', etc., are not logically incompatible with the existence of the type of laws the determinist claims may exist; if they do exist, it may not follow that statements of the kind quoted are always false, for it may be that these statements are true given certain conditions, which need not include the non-existence of any such laws.

Here, however, I shall not attempt to carry further any such inquiries into the meaning of these expressions or to press the view I have urged elsewhere, that the expression 'voluntary action' is best understood as excluding the presence of the various excuses. So I will not deal here with a determinist who is so incautious as to say that it may be false that anyone has ever acted 'voluntarily', 'of his own free will', or 'could have done otherwise than he did'. It will, I think, help to clarify our conception of criminal responsibility if I confront a more cautious sceptic who, without committing himself as to the meaning of those expressions or their logical or linguistic dependence on, or independence of, the negation of those types of law to which the determinist refers, yet criticizes our allocation of responsibility by reference to excusing conditions. This more cautious determinist says that whatever the expressions 'voluntary', etc. may mean, unless we have reasonable grounds for thinking there are no such laws, the distinctions drawn by these expressions cannot be regarded as of any importance, and there can be neither reason nor justice in allowing punishment to depend on the presence or absence of excusing conditions.

2

In the criminal law of every modern state responsibility for serious crimes is excluded or 'diminished' by some of the conditions we have referred to as 'excusing conditions'. In Anglo-American criminal law this is the doctrine that a 'subjective element', or 'mens rea', is required for criminal responsibility, and it is because of this doctrine that a criminal trial may involve investigations into the sanity of the accused; into what he knew, believed, or foresaw; or into the questions whether or not he was subject to coercion by threats or provoked into passion, or was prevented by disease or transitory loss of consciousness from controlling the movements of his body or muscles. These matters come up under the heads known to lawyers as Mistake, Accident, Provocation, Duress, and Insanity, and are most clearly and dramatically exemplified when the charge is one of murder or manslaughter.

Though this general doctrine underlies the criminal law, no legal system in practice admits without qualification the principle that *all* criminal responsibility is excluded by *any* of the excusing conditions. In Anglo-American law this principle is qualified in two ways. First, our law admits crimes of 'strict liability'.[2] These are crimes where it is no defence to show that the accused, in spite of the exercise of proper care, was ignorant of the facts that made his act illegal. Here he is liable to punishment even though he did not intend to commit an act answering the definition of the crime. These are for the most part petty offences contravening statutes that require the

[2] For an illuminating discussion of strict liability, see the opinion of Justice Jackson in *Morisette* v. *United States* (1952) 342 U.S. 246; 96 L. ed. 288; 72 S. Ct. 240. Also Sayre, 'Public Welfare Offences', 33 C.L.R. 55; Hall, *Principles of Criminal Law* (Indianapolis: Bobbs-Merrill Co., 1947). Chap. X.

maintenance of standards in the manufacture of goods sold for consumption; e.g. a statute forbidding the sale of adulterated milk. Such offences are usually punishable with a fine and are sometimes said by jurists who object to strict liability not to be criminal in any 'real' sense. Secondly, even in regard to crimes where liability is not 'strict', so that mistake or accident rendering the accused's action *unintentional* would provide an excuse, many legal systems do not accept some of the other conditions we have listed as excluding liability to punishment. This is so for a variety of reasons.

For one thing, it is clear that not only lawyers but scientists and plain men differ as to the relevance of some excusing conditions, and this lack of agreement is usually expressed as a difference of view regarding what kind of factor limits the human *capacity* to control behaviour. Views so expressed have indeed changed with the advance of knowledge about the human mind. Perhaps most people are now persuaded that it is possible for a man to have volitional control of his muscles and also to know the physical character of his movements and their consequences for himself and others, and yet be *unable* to resist the urge or temptation to perform a certain act; yet many think this incapacity exists only if it is associated with well-marked physiological or neurological symptoms or independently definable psychological disturbances. And perhaps there are still some who hold a modified form of the Platonic doctrine that Virtue is Knowledge and believe that the possession of knowledge[3] (and muscular control) is *per se* a sufficient condition of the capacity to comply with the law.[4]

Another reason limiting the scope of the excusing conditions

[3] This view is often defended by the assertion that the mind is an 'integrated whole', so that if the capacity for self-control is absent, knowledge must also be absent. See Hall, op. cit., p. 524: 'Diseased volition does not exist apart from diseased intelligence'; see also reference to the 'integration theory,' Chap. XIV.

[4] English judges have taken different sides on the issue whether a man can be said to have 'lost control', and killed another while in that condition, if he knew what he was doing and killed his victim intentionally. See *Holmes* v. *D.P.P.* (1946), A.C. at 597 (Lord Simon) and *A.G. for Ceylon* v. *Kumarasinghege* v. *Don John Perera* (1953), A.C. 200 (Lord Goddard).

is difficulty of *proof*. Some of the mental elements involved are much easier to prove than others. It is relatively simple to show that an agent lacked, either generally or on a particular occasion, volitional muscular control; it is somewhat more difficult to show that he did not know certain facts about either present circumstances (e.g. that a gun was loaded) or the future (that a man would step into the line of fire); it is much more difficult to establish whether or not a person was deprived of 'self-control' by passion provoked by others, or by partial mental disease. As we consider these different cases not only do we reach much vaguer concepts, but we become progressively more dependent on the agent's own statements about himself, buttressed by inferences from 'common-sense' generalizations about human nature, such as that men are capable of self-control when confronted with an open till but not when confronted with a wife in adultery. The law is accordingly much more cautious in admitting 'defects of the will' than 'defect in knowledge' as qualifying or excluding criminal responsibility. Further difficulties of proof may cause a legal system to limit its inquiry into the agent's 'subjective condition' by asking what a 'reasonable man' would in the circumstances have known or foreseen, or by asking whether 'a reasonable man' in the circumstances would have been deprived (say, by provocation) of self-control; and the system may then impute to the agent such knowledge or foresight or control.[5]

For these practical reasons, no simple identification of the necessary mental subjective elements in responsibility with the full list of excusing conditions can be made; and in all systems far greater prominence is given to the more easily provable elements of volitional control of muscular movement and knowledge of circumstances or consequences than to the other more elusive elements.

Hence it is true that legal recognition of the importance of excusing conditions is never unqualified; the law, like every

[5] See for a defence of the 'reasonable man' test (in cases of alleged provocation) *Royal Commission on Capital Punishment*, pp. 51–53 (paras. 139–45). This defence is not confined to the difficulties of proof.

other human institution, has to compromise with other values besides whatever values are incorporated in the recognition of some conditions as excusing. Sometimes, of course, it is not clear, when 'strict liability' is imposed, what value (social welfare?) is triumphant, and there has consequently been much criticism of this as an odious and useless departure from proper principles of liability.

Modern systems of law are however also concerned with most of the conditions we have listed as excusing conditions in another way. Besides the criminal law that requires men to do or abstain from certain actions whether they wish to or not, all such systems contain rules of a different type that provide legal facilities whereby individuals can give effect to their wishes by entering into certain transactions that alter their own and/or others' legal position (rights, duties, status, &c.). Examples of these civil transactions (acts in the law, *Rechtsgeschäfte*) are wills, contracts, gifts, marriage. If a legal system did not provide facilities allowing individuals to give legal effect to their choices in such areas of conduct, it would fail to make one of the law's most distinctive and valuable contributions to social life. But here too most of the mental conditions we have mentioned are recognized by the law as important not primarily as *excusing* conditions but as *invalidating* conditions. Thus a will, a gift, a marriage, and (subject to many complex exceptions) a contract may be invalid if the party concerned was insane, mistaken about the legal character of the transaction or some 'essential' term of it, or if he was subject to duress, coercion, or the undue influence of other persons. These are the obvious analogues of mistake, accident, coercion, duress, insanity, which are admitted by criminal law as excusing conditions. Analogously, the recognition of such conditions as invalidating civil transactions is qualified or limited by other principles. Those who enter in good faith into bilateral transactions of the kind mentioned with persons who appear normal (i.e. not subject to any of the relevant invalidating conditions) must be protected, as must third parties who may have purchased interests originating from a transaction that on the face of it seemed normal. Hence a technique has been introduced to safeguard such persons. This includes principles precluding, say, a party who has entered into a transaction by some mistake, from making this the basis of his defence against one who honestly took his words at face value and justifiably relied on them; there are also distinctions between transactions wholly invalidated *ab initio* (void) and those that are valid until denounced (voidable), to protect those who have relied on the transaction's normal form.

3

The similarity between the law's insistence on certain mental elements for both criminal responsibility and the validity of acts in the law is clear. Why, then, do we value a system of social control that takes mental conditions into account? Let us start with criminal law and its excusing conditions. What is so precious in its attention to these, and what would be lost if it gave this up? What precisely is the ground of our dissatisfaction with 'strict liability' in criminal law? To these fundamental questions, there still are, curiously enough, many quite discordant answers, and I propose to consider two of them before suggesting an answer that would stress the analogy with civil transactions.

The first general answer takes this form. It is said that the importance of excusing conditions in criminal responsibility is derivative, and it derives from the more fundamental requirement that for criminal responsibility there must be 'moral culpability', which would not exist where the excusing conditions are present On this view the maxim *actus non est reus nisi mens sit rea* refers to a morally evil mind Certainly traces of this view are to be found in scattered observations of English and American judges—in phrases such as 'an evil mind with regard to that which he is doing', 'a bad mind', or references to acts done not 'merely unguardedly or accidentally, without any evil mind.'[6]

[6] Lord Esher in *Lee* v. *Dangar*, (1892) 2 Q.B. 337.

Some of these well-known formulations were perhaps careless statements of the quite different principle that *mens rea* is an intention to commit an act that is wrong in the sense of legally forbidden. But the same view has been reasserted in general terms in England by Lord Justice Denning: 'In order that an act should be punishable, it must be morally blameworthy. It must be a sin.'[7] Most English lawyers would however now agree with Sir James Fitzjames Stephen that the expression *mens rea* is unfortunate, though too firmly established to be expelled, just because it misleadingly suggests that, in general, moral culpability is essential to a crime, and they would assent to the criticism expressed by a later judge that the true translation of *mens rea* is 'an intention to do the act which is made penal by statute or by the common law'.[8] Yet, in spite of this, the view has been argued by a distinguished American contemporary writer on criminal law, Professor Jerome Hall, in his important and illuminating *Principles of Criminal Law*, that *moral* culpability is the basis of responsibility in crime. Again and again in Chapters V and VI of his book Professor Hall asserts that, though the goodness or badness of the *motive* with which a crime is committed may not be relevant, the general principle of liability, except of course where liability is unfortunately 'strict' and so any mental element must be disregarded, is the '*intentional or reckless doing of a morally wrong act*'.[9] This is declared to be the essential meaning of *mens rea:* though *mens rea* differs in different crimes there is one 'common, essential element, namely, the voluntary doing of a morally wrong act forbidden by penal law'.[10] On this view the law inquires into the mind in criminal cases in order to secure that no one shall be punished in the absence of the basic condition of *moral* culpability. For it is just only to 'punish those who have intentionally committed moral wrongs, proscribed by law'.[11]

[7] Denning, *The Changing Law* (London: Stevens, 1953), p. 112.

[8] *Allard* v. *Selfridge*, (1925) 1 K.B. at 137. (Shearman J.) This is quoted by Glanville Williams in *Criminal Law, The General Part* (2nd edn.), p. 31, n. 3, where the author comments that the judge should have added 'or recklessness'.

[9] Hall, op. cit., p. 149. [10] Ibid., p. 167. [11] Ibid., p. 166.

Now, if this theory were merely a theory as to what the criminal law of a good society should be, it would not be possible to refute it, for it represents a moral preference: namely that legal punishment should be administered only where a 'morally wrong' act has been done—though I think such plausibility as it would have even as an ideal is due to a confusion. But of course Professor Hall's doctrine does not fit any actual system of criminal law because in every such system there are necessarily many actions (quite apart from the cases of 'strict liability') that if voluntarily done are criminally punishable, although our moral code may be either silent as to their moral quality, or divided. Very many offences are created by legislation designed to give effect to a particular economic scheme (e.g. a state monopoly of road or rail transport), the utility or moral character of which may be genuinely in dispute. An offender against such legislation can hardly be said to be morally guilty or to have intentionally committed a moral wrong, still less 'a sin' proscribed by law;[12] yet if he has broken such laws 'voluntarily' (to use Professor Hall's expression), which in practice means that he was not in any of the excusing conditions, the requirements of *justice* are surely satisfied. Doubts about the justice of the punishment would begin only if he were punished even though he was at the time of the action in one of the excusing conditions; for what is essential is that the offender, if he is to be *fairly* punished, must have acted 'voluntarily', and not that he must have committed some moral offence. In addition to such requirements of justice in the individual case, there is of course, as we shall see, a different type of requirement as to the *general* character of the laws.

It is important to see what has led Professor Hall and others to the conclusion that the basis of criminal responsibility *must*

[12] 'The criminal quality of an act . . . [cannot be] discovered by reference to any standard but one: Is the act prohibited with penal consequences? Morality and criminality are far from co-extensive; nor is the sphere of criminality necessarily part of a more extensive field covered by morality unless the moral code necessarily disapproves all acts prohibited by the State, in which case the argument moves in a circle.' Lord Atkin *Proprietory Articles Trade Association* v. *A.G. for Canada* (1931) A.C.310 at 324.

be moral culpability ('the voluntary doing of a morally wrong act'), for latent in this position, I think, is a false dilemma. The false dilemma is that criminal liability *must* either be 'strict'—that is, based on nothing more than the outward conduct of the accused—or *must* be based on moral culpability. On this view there is no third alternative and so there can be no reason for inquiring into the state of mind of the accused—'inner facts', as Professor Hall terms them—except for the purpose of establishing *moral* guilt. To be understood all theories should be examined in the context of argument in which they are advanced, and it is important to notice that Professor Hall's doctrine was developed mainly by way of criticism of the so-called objective theory of liability, which was developed, though not very consistently, by Chief Justice Holmes in his famous essays on common law.[13] Holmes asserted that the law did not consider, and need not consider, in administering punishment what in fact the accused intended but that it imputed to him the intention that an 'ordinary man', equipped with ordinary knowledge, would be taken to have had in acting as the accused did. Holmes in advocating this theory of 'objective liability' used the phrase 'inner facts' and frequently stressed that *mens rea*, in the sense of the actual wickedness of the party, was unnecessary. So he often identified 'mental facts' with moral guilt and also identified the notion of an objective standard of liability with the rejection of *moral* culpability as a basis of liability. This terminology was pregnant with confusion. It fatally suggests that there are only two alternatives: to consider the mental condition of the accused only to find moral culpability or not to consider it at all. But we are not impaled on the horns of any such dilemma: there are independent reasons, apart from the question of moral guilt, why a legal system should require a voluntary action as a condition of responsibility. These reasons I shall develop in a moment and merely summarize here by saying that the principle (1) that it is unfair and unjust to punish those who have not 'voluntarily' broken the law is a moral

[13] Holmes, *The Common Law*, Lecture II, 'The Criminal Law'.

principle quite distinct from the assertion (2) that it is wrong to punish those who have not 'voluntarily committed a moral wrong proscribed by law'.

The confusion that suggests the false dilemma—either 'objective' standards of liability or liability based on the 'inner fact' of *moral* guilt—is, I think, this. We would all agree that unless a legal system was as a whole morally defensible, so that its existence was better than the chaos of its collapse, and more good than evil was secured by maintaining and enforcing laws in general, these laws should not be enforced, and no one should be punished for breaking them. It *seems* therefore to follow, but does not, that we should not punish anyone unless in breaking the law he has done something morally wrong; for it looks as if the mere fact that a law has been voluntarily broken is not enough to justify punishment and the extra element required is 'moral culpability', at least in the sense that he should have done something morally wrong. What we need to escape confusion here is a distinction between two sets of questions. The first is a general question about the moral value of the laws: Will enforcing them produce more good than evil? If it does, then it is morally permissible to enforce them by punishing those who have broken them, unless in any given case there is some 'excuse'. The second is a particular question concerning individual cases: Is it right or just to punish this particular person? Is he to be excused on account of his mental condition because it would be unjust in view of his lack of knowledge or control—to punish him? The first, general question with regard to each law, is a question for the legislature; the second, arising in particular cases, is for the judge. And the question of responsibility arises only at the judicial stage. One necessary condition of the just application of a punishment is normally expressed by saying that the agent 'could have helped' doing what he did, and hence the need to inquire into the 'inner facts' is dictated not by the moral principle that only the doing of an *immoral* act may be legally punished, but by the moral principle that no one should be punished who could not help doing what he did. This is a necessary condition (unless strict liability is admitted) for the moral propriety of legal punishment and no doubt also for moral censure; in this respect law and morals are similar. But this similarity as to the one essential condition that there must be a 'voluntary' action if legal punishment or moral censure is to be morally permissible does not mean that legal punishment is morally permissible only where the agent has done something morally wrong. I think that the use of the word 'fault' in juristic discussion to designate the requirement that liability be excluded by excusing conditions may have blurred the important distinction between the assertions that (1) it is morally permissible to punish only voluntary actions and (2) it is morally permissible to punish only voluntary commission of a moral wrong.

4

Let me now turn to a second explanation of the law's concern with the 'inner facts' of mental life as a condition of responsibility. This is a Benthamite theory that I shall name the 'economy of threats' and is the contention that the required conditions of responsibility—e.g. that the agent knew what he was doing, was not subject to gross coercion or duress, was not mad or a small child—are simply the conditions that must be satisfied if the threat to punish announced by the criminal law is to have any effect and if the system is to be efficient in securing the maintenance of law at the least cost in pain. This theory is stated most clearly by Bentham; it is also to be found in Austin and in the report of the great Criminal Law Commission of 1833 of which he was a member. In a refined form it is implicit in many contemporary attempted 'dissolutions' of the problem of free will. Many accept this view as a common-sense utilitarian explanation of the importance that we attribute to excusing conditions. It appeals most to the utilitarian and to the determinist and it is interesting to find that Professor Glanville Williams in his recent admirable work on the general principles of criminal law,[14] when he wished to explain the exemption of the insane from legal responsibility compatibly with 'determinism', did so by reference to this theory.

Yet the doctrine is an incoherent one at certain points, I think, and a departure from, rather than an elucidation of, the moral insistence that criminal liability should generally be conditional on the absence of excusing conditions. Bentham's best statement of the theory is in Chapter XIII of his *Principles of Morals and Legislation:* 'Cases in Which Punishment Must be Inefficacious'. The cases he lists, besides those where the law is made *ex post facto* or not adequately promulgated, fall into two main classes. The first class consists of cases in which the penal threat of punishment could not prevent a person from performing an action forbidden by the law *or any action of the same sort*; these are the cases of infancy and insanity in which the agent, according to Bentham, has not the 'state or disposition of mind on which the prospect of evils so distant as those which are held forth by the law' has the effect of influencing his conduct. The second class consists of cases in which the law's threat could not have had any effect on the agent in relation to the *particular* act committed because of his lack of knowledge or control. What is wrong in punishing a man under both these types of mental conditions is that the punishment is wasteful; suffering is caused to the accused who is punished in circumstances where it could do no good.

In discussing the defence of insanity Professor Glanville Williams applies this theory in a way that brings out its consistency not only with a wholly utilitarian outlook on punishment but with determinism.

> For mankind in the mass it is impossible to tell whom the threat of punishment will restrain and whom it will not. For most [it] will succeed; for some it will fail, and the punishment must then be applied to these criminals in order to maintain the threat for persons generally. Mentally deranged persons, however, can be separated from the mass of mankind by scientific tests . . . Being a defined class their segregation from punishment does not impair the efficacy of the sanction for people generally.[15]

[14] Williams, op. cit. (1st edn.), pp. 346–7.

[15] Williams, loc. cit. This passage is, however, omitted from the 2nd edition of *Criminal Law, The General Part* (1961).

The point made here is that, if, for example, the mentally deranged (scientifically tested) are exempted, criminals will not be able to exploit this exemption to free themselves from liability, since they cannot bring themselves within its scope and so will not feel free to commit crimes with impunity. This is said in order to justify the exemption of the insane consistently with the tenet of determinism, in spite of the fact that from a determinist viewpoint

every impulse, if not in fact resisted, was in those circumstances irresistible. A so-called irresistible impulse is simply one in which the desire to perform a particular act is not influenced by other factors (like the threat of punishment). But on this definition every crime is the result of an irresistible impulse.

This theory is designed to fit not merely a utilitarian theory of punishment, but also the view that it is always false, if not senseless, to say that a criminal could have helped doing what he did. So on this theory, when we inquire into the mental state of the accused, we do not do so to answer the question, Could he help it? Nor of course to answer the question, Could the threat of punishment have been effective in his case?—for we know that it was not. The theory presents us with a far simpler conceptual scheme for dealing with the whole matter, since it does not involve the seemingly counter-factual speculation regarding what the accused 'could have done'. On this theory we inquire into the state of mind of the accused simply to find out whether he belongs to a defined class of persons whose exemption from punishment, if allowed, will not weaken the effect on others of the general threat of punishment made by the law. So there is no question of its being unjust or unfair to punish a particular criminal or to exempt him from punishment. Once the crime has been committed, the decision to punish or not has nothing to do with any moral claim or right of the criminal to have the features of his case considered, but only with the causal efficacy of his punishment on others. On this view the rationale of excuses is not (to put it shortly) that the accused should, in view of his mental condition, be excused whatever the effect of this on others; it is rather the mere fact that excusing him will not harm society by reducing the efficacy of the law's threats for others. So the criminal's mental condition is relevant simply to the question of the effect on others of his punishment or exemption.

This is certainly paradoxical enough. It seems to destroy the entire notion that in punishing we must be just to the particular criminal in front of us and that the purpose of excusing conditions is to protect him from society's claims. But, apart from paradox, the doctrine that we consider the state of a man's mind only to see if punishment is required in order to maintain the efficacy of threats for others is vitiated by a *non sequitur*. Before a man does a criminal action we may know that he is in such a condition that the threats cannot operate on him, either because of some temporary condition or because of a disease; but it does not follow—because the *threat* of punishment in his case, and in the case of others like him, is useless—that his *punishment* in the sense of the official administration of penalties will also be unnecessary to maintain the efficacy of threats for others at its highest. It may very well be that, if the law contained no explicit exemptions from responsibility on the score of ignorance, accident, mistake, or insanity, many people who now take a chance in the hope that they will bring themselves, if discovered, within these exempting provisions would in fact be deterred. It is indeed a perfectly familiar fact that pleas of loss of consciousness or other abnormal mental states, or of the existence of some other excusing condition, are frequently and sometimes successfully advanced where there is no real basis for them, since the difficulties of disproof are often considerable. The uselessness of a *threat* against a given individual or class does not entail that the *punishment* of that individual or class cannot be required to maintain in the highest degree the efficacy of threats for others. It may in fact be the case that to make liability to punishment dependent on the absence of excusing conditions is the most efficient way of maintaining the laws with the least cost in pain. But it is not *obviously* or *necessarily* the case.

It is clear, I think, that if we were to base our views of criminal responsibility on the doctrine of the economy of threats, we should misrepresent altogether the character of our moral preference for a legal system that requires mental conditions of responsibility over a system of total strict liability, or entirely different methods of social control such as hypnosis, propaganda, or conditioning.

To make this intelligible we must cease to regard the law as merely a causal factor in human behaviour differing from others only in the fact that it produces its effect through the medium of the mind; for it is clear that we look on excusing conditions as something that *protects* the individual against the claims of the rest of society. Recognition of their excusing force may lead to a lower, not a higher, level of efficacy of threats; yet—and this is the point—we would not regard that as sufficient ground for abandoning this protection of the individual; or if we did, it would be with the recognition that we had sacrificed one principle to another; for more is at stake than the single principle of maintaining the laws at their most efficacious level. We must cease, therefore, to regard the law simply as a system of stimuli goading the individual by its threats into conformity. Instead I shall suggest a mercantile analogy. Consider the law not as a system of stimuli but as what might be termed a *choosing* system, in which individuals can find out, in general terms at least, the costs they have to pay if they act in certain ways. This done, let us ask what value this system would have in social life and why we should regret its absence. I do not of course mean to suggest that it is a matter of indifference whether we obey the law or break it and pay the penalty. Punishment *is* different from a mere 'tax on a course of conduct'. What I do mean is that the conception of the law simply as goading individuals into desired courses of behaviour is inadequate and misleading; what a legal system that makes liability generally depend on excusing conditions does is to guide individuals' choices as to behaviour by presenting them with reasons for exercising choice in the direction of obedience, but leaving them to choose.

It is at this point that I would stress the analogy between the mental conditions that excuse from criminal responsibility and the mental conditions that are regarded as invalidating civil transactions such as wills, gifts, contracts, marriages, and the like. These institutions provide individuals with two inestimable advantages in relation to those areas of conduct they cover. These are (1) the advantage to the individual of determining by his choice what the future shall be and (2) the advantage of being able to predict what the future will be. For these institutions enable the individual (1) to bring into operation the coercive forces of the law so that those legal arrangements he has chosen shall be carried into effect and (2) to plan the rest of his life with certainty or at least the confidence (in a legal system that is working normally) that the arrangements he has made will in fact be carried out. By these devices the individual's choice is brought into the legal system and allowed to determine its future operations in various areas thereby giving him a type of indirect coercive control over, and a power to foresee the development of, official life. This he would not have 'naturally'; that is, apart from these legal institutions.

In brief, the function of these institutions of private law is to render effective the individual's preferences in certain areas. It is therefore clear why in this sphere the law treats the mental factors of, say, mistake, ignorance of the nature of the transaction, coercion, undue influence, or insanity as invalidating such civil transactions. For a transaction entered into under such conditions will not represent a real choice: the individual, might have chosen one course of events and by the transaction procured another (cases of mistake, ignorance, etc.), or he might have chosen to enter the transaction without coolly and calmly thinking out what he wanted (undue influence), or he might have been subjected to the threats of another who had imposed *his* choices (coercion).

To see the value of such institutions in rendering effective the individual's considered and informed choices as to what on the whole shall happen, we have but to conduct the experiment of imagining their absence: a system where no mental conditions would be recognized as invalidating such transactions and the consequent loss of control over the future that the individual would suffer. That such institutions *do* render individual choices effective and increase the powers of individuals to predict the course of events is simply a matter of empirical fact, and no form of 'determinism', of course, can show this to be false or illusory. If a man makes a will to which the law gives effect after his death, this is not, of course, merely a case of *post hoc:* we have enough empirical evidence to show that this was an instance of a regularity sufficient to have enabled us to predict the outcome with reasonable probability, at least in some cases, and to justify us, therefore, in interpreting this outcome as a consequence of making the will. There is no reason why we should not describe the situation as one where the testator *caused* the outcome of the distribution made. Of course the testator's choice in this example is only one prominent member of a complex set of conditions, of which all the other members were as necessary for the production of the outcome as his choice. Science may indeed show (1) that this set of conditions also includes conditions of which we are at the present moment quite ignorant and (2) that the testator's choice itself was the outcome of some set of jointly sufficient conditions of which we have no present knowledge. Yet neither of these two suppositions, even if they were verified, would make it false to say that the individuals' choice did determine the results, or make illusory the satisfaction got (*a*) from the knowledge that this kind of thing is possible, (*b*) from the exercise of such choice. And if determinism does not entail that satisfactions (*a*) or (*b*) are illusory, I for one do not understand how it could affect the wisdom, justice, rationality, or morality of the system we are considering.

If with this in mind we turn back to criminal law and its excusing conditions, we can regard their function as a mechanism for similarly maximizing within the framework of coercive criminal law the efficacy of the individual's informed and considered choice in determining the future and also his power to predict that future. We must start, of course, with the need for criminal law and its sanctions as at least some check on behaviour that threatens society. This implies a belief that the criminal law's threats actually do diminish the frequency of antisocial behaviour, and no doubt this belief may be said to be based on inadequate evidence. However, we must clearly take it as our starting point: if this belief is wrong, it is so because of lack of empirical evidence and not because it contradicts any form of determinism. Then we can see that by attaching excusing conditions to criminal responsibility, we provide each individual with benefits he would not have if we made the system of criminal law operate on a basis of total 'strict liability'. First, we maximize the individual's power at any time to predict the likelihood that the sanctions of the criminal law will be applied to him. Secondly, we introduce the individual's choice as one of the operative factors determining whether or not these sanctions shall be applied to him. He can weigh the cost to him of obeying the law—and of sacrificing some satisfaction in order to obey—against obtaining that satisfaction at the cost of paying 'the penalty'. Thirdly, by adopting this system of attaching excusing conditions we provide that, if the sanctions of the criminal law are applied, the pains of punishment will for each individual represent the price of some satisfaction obtained from breach of law. This, of course, can sound like a very cold, if not immoral attitude toward the criminal law, general obedience to which we regard as an essential part of a decent social order. But this attitude seems repellent only if we assume that all criminal laws are ones whose operation we approve. To be realistic we must also think of bad and repressive criminal laws; in South Africa, Nazi Germany, Soviet Russia, and no doubt elsewhere, we might be thankful to have their badness mitigated by the fact that they fall only on those who have obtained a satisfaction from knowingly doing what they forbid.

Again, the value of these three factors can be realized if we conduct the *Gedankenexperiment* of imagining criminal law operating without excusing conditions. First, our power of predicting what will happen to us will be immeasurably diminished; the likelihood that I shall choose to do the forbidden act (e.g. strike someone) and so incur the sanctions of the criminal law may not be very easy to calculate even under our system: as a basis for this prediction we have indeed only the knowledge of our own character and some estimate of the temptations life is likely to offer us. But if we are also to be liable if we strike someone by accident, by mistake, under coercion, etc., the chances that we shall incur the sanctions are immeasurably increased. From our knowledge of the past career of our body considered as a *thing*, we cannot infer much as to the chances of its being brought into violent contact with another, and under a system that dispensed with the excusing condition of, say, accident (implying lack of intention) a collision alone would land us in jail. Secondly, our choice would condition what befalls us to a lesser extent. Thirdly, we should suffer sanctions without having obtained any satisfaction. Again, no form of determinism that I, at least, can construct can throw any doubt on, or show to be illusory, the real satisfaction that a system of criminal law incorporating excusing conditions provides for individuals in maximizing the effect of their choices within the framework of coercive law. The choices remain choices, the satisfactions remain satisfactions, and the consequences of choices remain the consequences of choices, even if choices are determined and even if other 'determinants' besides our choices condition the satisfaction arising from their being rendered effective in this way by the criminal law.

It is now important to contrast this view of excusing conditions with the Benthamite explanation, e.g. discussed in Part 3 of this paper. On that view excusing conditions were treated as conditions under which the laws' threat could operate with maximum efficacy. They were recognized *not* because they ensured justice to individuals considered separately, but because sanctions administered under those conditions were believed more effective and economical of pain in securing the general conformity to law. If these beliefs as to the *efficacy* of excusing conditions could be shown false, then all reasons for recognizing them as conditions of criminal responsibility would disappear. On the present view, which I advocate, excusing conditions are accepted as independent of the efficacy of the system of threats. Instead it is conceded that recognition of these conditions may, and probably does, diminish that efficacy by increasing the number of conditions for criminal liability and hence giving opportunities for pretence on the part of criminals, or mistakes on the part of tribunals.

On this view excusing conditions are accepted as something that may conflict with the social utility of the law's threats; they are regarded as of moral importance because they provide for all individuals alike the satisfactions of a choosing system. Recognition of excusing conditions is therefore seen as a matter of protection of the individual against the claims of society for the highest measure of protection from crime that can be obtained from a system of threats. In this way the criminal law respects the claims of the individual as such, or at least as a *choosing being*, and distributes its coercive sanctions in a way that reflects this respect for the individual. This surely is very central in the notion of justice and is *one*, though no doubt only one, among the many strands of principle that I think lie at the root of the preference for legal institutions conditioning liability by reference to excusing conditions.

I cannot, of course, by unearthing this principle, claim to have solved everyone's perplexities. In particular, I do not know what to say to a critic who urges that I have shown only that the system in which excusing conditions are recognized protects the individual better against the claims of society than one in which no recognition is accorded to these factors. This seems to me to be enough; yet I cannot satisfy the complaint, if he makes it, that I have not shown that we are justified in punishing anyone *ever*, at all, under any conditions. He may say that even the criminal who has committed his crime in the most deliberate and calculating way and has shown himself throughout his life competent in maximizing what he thinks his own interests will be little comforted when he is caught and punished for some major crime. At *that* stage he will get little satisfaction if it is pointed out to him (1) that he has obtained some satisfaction from his crime, (2) that he knew that it was likely he would be punished and that he had decided to pay for his satisfaction by exposing himself to this risk, and (3) that the system under which he is punished is not one of strict liability, is not one under which a man who accidentally did what he did would also have suffered the penalties of the law.

5

I will add four observations *ex abundante cautela*.

(i) The elucidation of the moral importance of the mental element in responsibility, and the moral odium of strict liability, that I have indicated, must not be mistaken for a psychological theory of motivation. It does not answer the question, Why do people obey the law? It does not assert that they obey only because they choose to obey rather than pay the cost. Instead, my theory answers the question, Why *should* we have a law with just these features? Human beings in the main do what the law requires without first choosing between the advantage and the cost of disobeying, and when they obey it is not usually from fear of the sanction. For most the sanction is important not because it inspires them with fear but because it offers a guarantee that the antisocial minority who would not otherwise obey will be coerced into obedience by fear. To obey without this assurance might, as Hobbes saw, be very foolish: it would be to risk going to the wall. However, the fact that only a few people, as things are, consider the question Shall I obey or pay?, does not in the least mean that the standing possibility of asking this question is unimportant: for it secures just those values for the individual that I have mentioned.

(ii) I must, of course, confront the objection which the Marxist might make, that the excusing conditions, or indeed *mutatis mutandis* the invalidating conditions, of civil transactions are of no use to many individuals in society whose economic or social position is such that the difference between a law of strict liability and a law that recognizes excusing conditions is of no importance.

It is quite true that the fact that criminal law recognizes excusing mental conditions may be of no importance to a person whose economic condition is such that he cannot profit from the difference between a law against theft that is strict and one that incorporates excusing conditions. If starvation 'forces' him to steal, the values the system respects and incorporates in excusing conditions are nothing to him. This is of course similar to the claim often made that the freedom that a political democracy of the Western type offers to its subjects is merely formal freedom, not real freedom, and leaves one free to starve. I regard this as a confusing way of putting what may be true under certain conditions: namely, that the freedoms the law offers may be *valueless* as playing no part in the happiness of persons who are too poor or weak to take advantage of them. The admission that the excusing condition may be of no value to those who are below a minimum level of economic prosperity may mean, of course, that we should incorporate as a further excusing condition the pressure of gross forms of economic necessity. This point, though possibly valid, does not seem to me to throw doubt on the principle lying behind such excusing conditions as we do recognize at present, nor to destroy their genuine value for those who are above the minimum level of economic prosperity; for the difference between a system of strict liability and our present system plays a part in their happiness.

(iii) The principle by reference to which I have explained the moral importance of excusing conditions may help to clarify an old dispute, apt to spring up between lawyers on the one hand and doctors and scientists on the other, about the moral basis of punishment.

From Plato to the present day there has been a recurrent insistence that if we were rational we would always look on crime as a disease and address ourselves to its cure. We would do this not only where a crime has actually been committed but where we find well-marked evidence that it will be. We would take the individual and treat him as a patient before the deed was done. Plato,[16] it will be remembered, thought it

[16] Plato, *Protagoras*, 324; *Laws*, 861, 865.

superstitious to look back and go into questions of responsibility or the previous history of a crime except when it might throw light on what was needed to cure the criminal.

Carried to its extreme, this doctrine is the programme of Erewhon, where those with criminal tendencies were sent by doctors for indefinite periods of cure; punishment was displaced by a concept of social hygiene. It is, I think, of some importance to realize why we should object to this point of view; for both those who defend it and those who attack it often assume that the *only* possible consistent alternative to Erewhon is a theory of punishment under which it is justified simply as a return for the moral evil attributable to the accused. Those opposed to the Erewhonian programme are apt to object that it disregards *moral* guilt as a necessary condition of a just punishment and thus leads to a condition in which any person may be sacrificed to the welfare of society. Those who defend an Erewhonian view think that their opponents' objection must entail adherence to the form of retributive punishment that regards punishment simply as a justified return for the moral evil in the criminal's action.

Both sides, I think, make a common mistake: there *is* a reason for making punishment conditional on the commission of crime and respecting excusing conditions, which is quite independent of the form of retributive theory that is often urged as the only alternative to Erewhon. Even if we regard the over-all purpose of punishment as that of protecting society by deterring persons from committing crimes and insist that the penalties we inflict be adapted to this end, we can in perfect consistency and with good reason insist that these punishments be applied only to those who have broken a law and to whom no excusing conditions apply. For this system will provide a measure of protection to individuals and will maximize their powers of prediction and the efficacy of their choices in the ways that I have mentioned. To see this we have only to ask ourselves what in terms of these values we should lose (however much else we might gain) if social hygiene and a *system of compulsory treatment* for those with detectable criminal tendencies were throughout substituted for our system of punishment modified by excusing conditions. Surely the realization of what would be lost, and not a retributive theory of punishment, is all that is required as a reason for refusing to make the descent into Erewhon.

(iv) Finally, what I have written concerns only *legal* responsibility and the rationale of excuses in a legal system in which there are organized, coercive sanctions. I do not think the same arguments can be used to defend *moral* responsibility from the determinist, if it is in any danger from that source.

III

MURDER AND THE PRINCIPLES OF PUNISHMENT: ENGLAND AND THE UNITED STATES[1]

I

ENGLISH people have probably been more disturbed and more divided by the use of the death penalty for murder than any people who still retain it as a form of punishment for that offence. Since Bentham ceased writing in 1832, the question of the death penalty has always been the subject of anxious scrutiny in England. The issue is considered to be one which deeply concerns a man's conscience, and it is recognized that views on this matter may cut across political loyalties. Accordingly, when the matter is debated in the House of Commons it has usually been thought right to relax the strict party discipline which normally governs debate and to allow members to vote free from claims of party loyalty. This has resulted in members of different parties voting on the same side when the issue of capital punishment has arisen. Outside the legislature, however, advocacy of abolition has seldom been left to the unorganized efforts of individuals. Associations for the abolition of the death penalty have for many years been a familiar feature of English life; some of these have commanded the services of distinguished writers and speakers, and received

[1] This article was written in 1957. Certain later changes in English and American law are described in the Notes, pp. 245-51 together with later comparative statistics. Except where the contrary is stated in the Notes the comparisons of murder and its treatment between England and the United States still hold good. For the period since 1964 see Appendix, p. 268 *infra*.

some financial support, and conducted studies into statistical and other relevant facts.[2]

In the last 100 years there have been successive Parliamentary assaults on the death penalty, but only in the last ten years have these had the support of an actual majority. As far back as 1866, only four years after the abolition of the death penalty for a wide range of offences, a Select Committee of twelve considered its abolition for the offence of murder and five members of this body voted for abolition. In 1930 a Select Committee of the House of Commons reported in favour of a suspension of the death penalty for murder in cases tried by civil courts for an experimental period of five years.[3] Since the war, concern about the use of the death penalty for murder has been much intensified and three times in the last ten years Parliament came very near to suspending it for all forms of murder. In 1948 the House of Commons voted for a five-year suspension, but this provision was deleted by the House of Lords from the great Criminal Justice Bill of that year. In 1955 a similar motion was defeated in the House of Commons by a vote of 245 to 214, a majority of only thirty-one. In February 1956 the House of Commons on a free vote passed a resolution calling for the abolition or suspension of the death penalty by a vote of 292 to 246, a majority of forty-six.

This resolution marks the crossing of a great divide in the English treatment of murder and its words bear repetition here:

[2] The Howard League for Penal Reform for twenty-five years prior to 1950 worked for the abolition of capital punishment and made the first collection of statistics from abolition countries in Scandinavia. This inquiry led to the formation of the National Council for the Abolition of the Death Penalty, which represented a number of national societies opposed to capital punishment and published a study made by its secretary of homicide rates in abolition countries. *Calvert, Capital Punishment in the Twentieth Century* (1927).

[3] See *Select Committee on Capital Punishment, Report* (1930). Six Conservative members of this committee of fifteen withdrew from the committee; the recommendations are those of the remaining majority of nine (seven Labour, two Liberal). This report was never debated in Parliament. See *Calvert, The Death Penalty Enquiry* (1931).

That this house believes that the death penalty for murder no longer accords with the needs or the true interests of a civilised society and calls upon Her Majesty's Government to introduce forthwith legislation for its abolition or for its suspension for an experimental period.[4]

In due course the House of Lords rejected the legislation which was passed by the House of Commons in the spirit of this resolution.[5] There are ways in our curious Constitution of circumventing the opposition of our Upper Chamber, but in practice these are not available unless the Government of the day is in favour of the measures which the House of Commons passes. In this case the Government was opposed to suspension or abolition of the death penalty for all forms of murder and refused to lend its aid. It is perhaps worth noting that it was by no means clear that the majority of the electorate concurred with the majority of the House of Commons. The Government was convinced that public opinion was opposed to the abolition of the death penalty, and certainly many members of Parliament must have voted for abolition even though they believed that a majority of their own constituents were opposed to it.[6] This illustrates the survival of the theory that the English Member of Parliament is not a delegate but

[4] 548 *H.C. Deb.* 2652, 2655 (1956).

[5] This was the bill introduced before the House of Commons resolution by a private member, Mr. Silverman, on 15 Nov. 1955, providing for the suspension of the death penalty for ten years to continue indefinitely after that period if no action was taken to restore it.

[6] See 548 *H.C. Deb.* 2575 (1956). 'I shall not argue that there is yet a majority of public opinion in favour of abolition.' (Mr. Herbert Morrison.) See also 198 *H.L. Deb.* 790, 804 (1956), for criticism of members of the Commons disregarding their constituents' known views. The issue was never put before the electorate at any general election and no member of the House of Commons who voted in favour of it referred to it in his election address; (Ibid. at 824). In the House of Lords debate of the Silverman Bill in July 1956, eight bishops voted for the suspension of the death penalty and two against; two peers holding judicial office voted for suspension and eight against; (Ibid. at 840–2). Opposition to abolition or suspension was particularly strong among the police and prison services. See *Royal Commission on Capital Punishment, Report*, Cmd. No. 8932, para. 61 (1953) (hereinafter cited as *Royal Commission Report*). A Gallup Poll in 1953 showed 73 per cent in favour of the death penalty and a poll in 1955 showed 50 per cent in favour and 37 per cent against. These fluctuations, however, were said to be influenced by two recent cases. 198 *H.L. Deb.* 689 (1956).

a representative of his constituents. His duty even in a democracy is not to act on some real or supposed mandate from his constituents to vote in a given manner, but, at least when freed from party discipline, to consider each measure as it comes up before the House of Commons and to vote in accordance with his judgement.

Though no legislation was passed to give effect to the resolution of February 1956, two things of major importance resulted from it. First, all executions were suspended. The method by which this suspension was secured was the granting of reprieves in exercise of the Royal Prerogative on the 'advice' of the Home Secretary. Until the new legislation mentioned below such a reprieve was granted in every case, though prior to February 1956 in more than half the number of cases where a prisoner was convicted of murder and sentenced to death the sentence was carried out.[7]

The second result of the vote in the House of Commons was that the Government itself introduced a compromise measure which became law under the title of the Homicide Act on 21 March 1957. This act eliminates the death penalty except for five categories of murder, the most important of which are murders done in the course of theft, murders by shooting or by causing an explosion, and murder of a police officer acting in the execution of his duty.[8]

In the intervals between these Parliamentary debates, public discussion of the death penalty was conducted vigorously in the press, on the radio, and at public meetings organized by various bodies, including the Howard League. Since 1953,

[7] Executions were resumed in July 1957.

[8] Ibid. s. 5. The other cases of capital murder under the act are: murder done in the course of or for the purpose of preventing lawful arrest or of effecting or assisting an escape from legal custody, murder of a prison officer by a prisoner, and repeated murders. The act also abolishes constructive malice and introduces into the law of murder the doctrine of diminished responsibility, which provides that a person who kills shall not be convicted of murder if he was suffering from such abnormality of mind as 'substantially impaired his mental responsibility', but only of manslaughter carrying a maximum penalty of imprisonment for life. Ibid. s. 2. For criticisms, see Elliott, 'The Homicide Act, 1957', *Crim. L. R.*, (1957) p. 282; Prevezer, 'The English Homicide Act: A New Attempt to Revise the Law of Murder,' 57 *C. L. R.* (1957), p. 624.

this discussion has been of a markedly higher quality than before. This was due to the publication, in 1953, and the subsequent wide dissemination of the *Report of the Royal Commission on Capital Punishment*,[9] summing up the results of four years' study of the facts, the figures, the law, and the moral principles which stand behind the law, in relation to murder and its punishment. This Commission visited many parts of Europe and the United States and addressed questionnaires to many countries in search of information; it was aided by evidence given to it by many celebrated experts and jurists. Among those in the United States were Justice Felix Frankfurter, Professor Herbert Wechsler of Columbia University and Professor Thorsten Sellin. Within the confines of this report there is a far more comprehensive, dispassionate, and lucid evaluation of the arguments both as to questions of fact and to questions of law and principle relevant to murder and its punishment, than in any of the many books published in either of our countries on this subject. Certainly the publication of this report in England introduced altogether new standards of clarity and relevance into discussions of a subject which had too often been obscured by ignorance and prejudice. The value of this most remarkable document was not diminished by the fact that the Commission's terms of reference postulated the retention of the death penalty and extended only to the consideration of the *limitations* on its use; nor was it diminished by the fact that the recent Homicide Act of 1957 proceeded on principles in two respects opposed to the conclusions of the Royal Commission.

2

It is profitable, I think, to consider some major contrasts between the way in which murder and its punishment is

[9] Cmd. 8932.

regarded in England and this country. Of course I am very conscious of the fallacy of speaking of the United States as if it were a single country; and there is certainly a great diversity in the statutory definition of murder and in its treatment in the different states of the Union.[10] We share indeed the common law concept of murder, malice aforethought, and the distinction between murder and manslaughter, but against this common background five major differences stand out.

First, until the Homicide Act of 1957 English law had never admitted the notion of different degrees of murder, but had adhered obstinately to the simple division of criminal homicide into murders for which the death sentence is mandatory and manslaughter for which only a maximum sentence of imprisonment for life is prescribed. Efforts for many years had been made to introduce the gradations of murder so familiar in American law. In 1866 a Select Commission reported in favour of dividing murder into two degrees, for one of which capital punishment was reserved, and two successive Governments and several private members introduced bills to give effect to this recommendation, but without success.[11] In 1948, after the House of Lords had deleted the suspension of the death penalty from the Criminal Justice Bill, the House of Commons voted for a compromise clause introduced by the Government, which reserved the death penalty for five categories of murder committed with express malice, but this was rejected by the House of Lords and dropped from the bill.[12]

Finally the Royal Commission, in its report of 1953, gave

[10] Ten states employ the common-law definition without grading into degrees. Capital punishment has been abolished in Maine, Michigan, Wisconsin, and Minnesota, and, except for murders committed by those already under sentence of life imprisonment, in Rhode Island and North Dakota.

Where the death penalty exists, a discretion is usually given to the court or jury to substitute life imprisonment, but the death penalty is mandatory for first degree murder in Connecticut, Massachusetts, North Carolina, and Vermont, and, except for most felony murder, New York.

[11] The government introduced bills in 1866 and 1867. See *Royal Commission Report*, pp. 467–70, App. 12.

[12] See ibid. pp. 170–2, 471.

more careful consideration to the introduction of degrees of murder in English law than this subject had ever received before; yet, after a most exhaustive consideration of the practice in the United States and elsewhere, the conclusion was reached that the quest for a satisfactory definition of degrees of murder was 'chimerical' and must be abandoned.[13] Only in 1957 was a breach made in this tradition by the Homicide Act, which reserves the death penalty for five classes of murder. The particular classes chosen may appear somewhat curious to American lawyers, but it is to be remembered that they do not represent an attempt to distinguish between murders according to heinousness or moral gravity, but to select for capital punishment those types of murder in which the deterrent effect is likely to be most powerful.

Of greater importance than the English refusal to contemplate degrees of murder is the even greater reluctance of English lawyers to confer a discretion upon the court as to the penalty in murder cases and, above all, the solid conviction that such a discretion should not be imparted to a jury. Although, as Justice Frankfurter has stated, the various American states present a crazy quilt pattern defeating any generalization, the *normal* method by which American justice seeks to determine the appropriate penalty in murder cases is a combination of degrees of murder and the device of entrusting a discretion as to the penalty to the court, except for certain cases where the death penalty is mandatory for first-degree murder. English legal and public opinion has, of course, always been disturbed by the fact that the rigid English law of murder makes the death penalty mandatory for offences of widely different moral character; and indeed the death penalty would not have been tolerated at all had it been carried out in all cases where it was imposed. But until very recently the method of mitigating the rigidity of the law has been the clumsy device of leaving the ultimate disposition of each case to the Executive. Accordingly, it is the Home Secretary,

[13] Ibid. p. 189 (para. 534).

exercising the Royal Prerogative of mercy, who in the end draws the distinction so universally felt between, e.g. the cold-blooded murderer out for gain and the woman who kills an imbecile child to whom she can no longer attend. Over the last fifty years, reprieves have been granted in nearly half of the cases where the courts have sentenced the prisoner to death.

Thus it has been said with some truth that the English courts merely determine which murderers *may* be executed.[14] Perhaps English lawyers have never thought out in detail their reasons for entrusting to a single man, in the person of the Home Secretary, a discretion which they would withhold from the court and above all from the jury. However, the unwillingness to entrust this discretion to the court, where it is more than mere conservatism, springs from the knowledge that the Home Secretary, in reviewing a case and considering a reprieve, has access to a wide range of information which could not come to light in court and goes far beyond the facts required to show guilt at trial. He may consider the whole background of a prisoner's life, including information which has come to light only since the trial, and, odious as his responsibility is felt to be, it is the conviction of most English lawyers that he does something which the courts cannot do.

On the other hand, the disadvantages of this system have been felt acutely in recent years. The Royal Commission felt that the main evil of the existing law was that a grotesque combination of a solemn sentence of death passed after a trial in court, followed, in nearly half the cases, by a reprieve made by the Executive, was needed to achieve a morally tolerable result. This clumsy expedient, whatever its advantages, could scarcely fail to create the impression that the law which the courts administer lags behind the best informed, enlightened, standards of the day, and almost as often as not the decision reached by the court had to be set right by the

[14] Michael and Wechsler, 'A Rationale of the Law of Homicide', 37 *C. L. R.* (1937), p. 701, at 706 n. 19.

Executive.[15] The Royal Commission reached the conclusion that if the death penalty were not to be abolished for murder the only satisfactory solution was to entrust a discretion to the jury to enable it to give effect to considerations which at present lead the Home Secretary to recommend a reprieve.[16] In England this conclusion has not been adopted; 'jury discretion' is too generally distrusted as an expedient for determining punishment.[17]

The third major difference between the law in England and in the United States is in relation to felony murder, or, as we call it, 'constructive murder'. This has finally been eliminated from English law by the Homicide Act 1957 in deference to the conviction that it is unjust or in some way inconsistent with enlightened principles of punishment that a person should be convicted of murder if he killed a person while engaged upon some felonious act not in itself likely to result in the death of, or grievous injury to, another person.[18] Gradually the scope of felony murder had been reduced, first by the insistence that it should be used only in cases where actual violence had in fact been the cause of death, and then by the insistence that the death should have been the natural or probable outcome of the felony upon which the prisoner was engaged. Though there had been a few cases in recent years where a broader interpretation of felony murder was given by an English court,[19] on the whole the conviction was widespread that it had ceased to be defensible.

This, of course, is in clear contrast with the United States where it is not deemed extraordinary to insist that a man who

[15] See *Royal Commission Report*, paras. 17–22, 606–8, 790 ss.2 and 42.

[16] Ibid. paras. 594–611, 790 ss.6–43.

[17] See especially the debate in the House of Lords on the scheme for 'jury discretion' 185 H.L. Deb. 137–88 (1953). All the legal members present except Lord Chorley (ibid. 170) condemned it as completely unworkable and the Lord Chief Justice said 'Rather than take part in such a performance as that I would resign the office I hold, for *I think it would be destructive of everything in British law*'; ibid. 177 (my italics).

[18] See *Royal Commission Report* pp. 34–41.

[19] e.g. *R.* v *Jarmain* (1946) K.B. 74.

embarks on violent crime and sets a chain of events in motion must 'take the consequences'. This attitude, which conflicts with the importance attached to the principle that a 'subjective test' of criminal liability should be adhered to as far as possible,[20] no doubt reflects the wide prevalence of crimes of violence and the common fear of rape in some parts of America. In England, even judges regarded as generally conservative in their attitudes toward the law of murder, such as the Lord Chief Justice and Mr. Justice Humphreys, concurred in the view that constructive malice should be abolished and that murder should be confined in terms or in effect to cases of intentional killing or infliction of grievous bodily harm.

The fourth difference which requires attention is the attitude, common in English penological thought, that a sentence of imprisonment longer than ten years should not be served except under the most extreme circumstances. Under the practice prior to the Homicide Act 1957, a murderer whose death sentence had been commuted for one of life imprisonment very rarely served a period of more than fifteen years, and the usual period served was very much less.[21] By contrast, in some states in America sentences of twenty-eight years are not uncommonly served for first-degree murder and seventeen years for second-degree murder.[22] It is of course obvious that, since murderers who would have previously been executed will now be sentenced to imprisonment as a result of the Homicide Act, a new consideration will have to be given to the question of the length of imprisonment which can be imposed consistently with our general penological notions. Much experience in abolition countries on

[20] Still 'objective standards' creep in by the back door. See *R.* v. *Ward* (1956) 1 Q.B. 351 (C.C.A.), criticized by Prevezer, 'Murder by Mistake', *Crim. L. R.* (1956), p. 375.

[21] *Royal Commission Report*, paras. 644–656 (But see notes *infra* p. 250).

[22] Ibid., App. 16, Table B:

Massachusetts 1900–50		
1st. degree murder	Average 28 yrs.	Longest 41 yrs.
2nd. degree murder	Average 17 yrs.	Longest 40 yrs.

the continent is available; it is thought by many to demonstrate that even where there is no death penalty, extended sentences of imprisonment are not in fact necessary for the public safety. Only one of the murderers whose sentence of death had been commuted to imprisonment and had been released is known to have committed a murder in England in this century.

Last, we should note the attitude of English law in regard to the notion of double jeopardy in criminal cases. If some substantial error is made by the trial court, whether it be a misdirection to the jury or the admission of inadmissible evidence, the result in England is that the conviction is quashed by the court of appeal and the prisoner acquitted. In such circumstances the prisoner does not run the risk of a second trial, and in very many cases where a prisoner in the United States would be subjected to a second trial he would, in England, be acquitted on appeal.[23] There is indeed a power given to our court of criminal appeal to maintain the decision of the trial court in spite of misdirection or other error, but only if such error is not a 'substantial miscarriage of justice'.[24] It might be thought that the result of the English system is that the appellate courts are less quick to detect error in the trial than is true in the United States, but I can only say, after consideration of many cases, that I think that this is not so. The result of our system is that we never have any instances of considerable delay between the actual conviction and sentence and its execution. A period of more than three months would be considered indecent and an intolerable cruelty to the prisoner and his family. This, of course, contrasts with the

[23] See *Woolmington* v. *Director of Public Prosecution*, (1935) A.C. 462. The trial judge wrongly directed the jury that if the fact of killing by the accused was established by the prosecution, it was then for the accused to show that he had not killed with malice aforethought. On appeal the conviction was quashed because of this misdirection, and the prisoner was acquitted and discharged.

[24] Criminal Appeal Act, 1907, 7 Edw. VII, c. 23, s. 4(1). Only where 'if the jury had been properly directed they would inevitably come to the same conclusion' can misdirection be treated as other than a substantial miscarriage of justice for the purpose of this section. See *Woolmington* v. *Director of Public Prosecution, supra*, at p. 482.

occasional case in the United States where years may intervene between sentence of death and its execution.

3

So far we have considered the chief differences in the law and penological ideas concerning murder in our two countries. Let us now turn to the question of the rate and types of murder prevalent in England and in the United States. Here we do indeed tread upon ground which is full of pitfalls. We must remember how blunt are our sociological tools for assessing the quantity of crime of any type, and in particular the crime of homicide. Of course, it is easy to find out how many are charged and convicted of specific crimes; but, if we are to begin the possibly hopeless task of assessing the value of any particular punishment, what is needed is some rational estimate of the underlying figures of the crimes actually perpetrated.

In England, for more than seventy years, figures showing the number of 'murders known to the police' have, along with similar figures for other crimes, been collected by the Government from all of the various police forces in the country. These are annually presented to Parliament by the Home Office and published as the *Criminal Statistics for England and Wales*. The authorities are well aware of the difficulties of classifying particular cases and have often emphasized that, while every effort is made to secure uniformity as between different police forces, and scrupulous attention is paid to the need to distinguish cases where a reported death might be due to suicide, self-defence, accident, and felonious attack, it is impossible to guarantee that any figures so collected are immune from error.

Figures compiled in this way must necessarily be imperfect as a guide to the amount of murder, if only because there must be a certain number of cases which are never reported to the police. Yet they are considered to be the best index; they are much better, for example, than figures from registration offices[25] showing the number of deaths officially recorded as due to murder. Hence it is regrettable that until recently only Great Britain, the Commonwealth Countries, Denmark, and a few states in the United States kept full figures on this basis. Since 1930, however, the *Uniform Crime Reports*, issued biannually by the F.B.I., show figures obtained from police authorities in most of the urban and rural areas of the country for cases of 'murder and non-negligent manslaughter' known to the police. These figures now cover 90 per cent of the urban population, 68 per cent of the rural population, and 81 per cent of the total population.[26] They have been fiercely criticized at times[27] and, with respect to certain crimes at any rate, it is clear that the basis of classification may have varied from district to district and perhaps was influenced by the wish to make things look better than they are. A warning is now contained in these reports to the effect that, in publishing the data sent in by the chiefs of police of different cities, the F. B. I. does not vouch for their accuracy. Nevertheless it seems that the general standards for reporting crimes have risen considerably,[28] and, though they must be open to question at many points, it is now reasonable to regard these figures as a rough minimum estimate of the homicides that they report. Of course, if there is a need for caution in comparing the figures reported by the police for one district with the figures reported by the police for another even in the same country, obviously there is an even greater need for caution in comparing the figures reported by the police of different countries. Yet, with all these qualifications, it is still illuminating to draw attention

[25] For a criticism of such figures see *Royal Commission Report,* App. 6, para. 27.

[26] See, e.g. 26 *U.S. Dept of Justice, Uniform Crime Report,* (1955,) p. 72 (hereinafter cited as *Uniform Crime Reports*).

[27] See Warner, 'Crimes Known to the Police—An Index of Crime?' 45 *Harv. L. Rev.* (1932), p. 307.

[28] See 25 *Uniform Crime Reports* 72–73 (1954) where the claim is made that the 'reliability of major crime estimates is considered excellent'. A short account is also given of the methods now used to correct deviations from 'acceptable standards' in record keeping and reporting and of several methods used for test checking of reported figures.

to some contrasts between the figures for Great Britain and those for the United States.

The relevant figures for England and Wales most worth attention are, I think, the following. In the fifty years from 1900 to 1949 inclusive, a total of 7,454 murders were known to the police; this included 2,001 babies under the age of one year.[29] If we break down these fifty years into five decades of ten years the rate per million of population of murder (including babies under one year) is an average rate of 4.6, 4.1, 3.9, 3.3, 4.0 for each of these five decades.[30] The annual average figure of murder known to the police for these fifty years is 149 to the nearest unit; and the actual figure for each of these fifty years is quite frequently very near this average.

In this same period of fifty years in England and Wales, 1,210 persons were sentenced to death for murder but only 632 were executed. This is an annual average of twenty-four death sentences and thirteen executions.[31] Slightly more than half the number of those sentenced to death were executed and the remainder are accounted for almost wholly by the intervention of the Executive.

In rough figures, therefore, in England and Wales during this period one person was executed for every twelve murders known to the police, and one person was convicted for every six murders known to the police.[32]

[29] *Royal Commission Report*, App. 3, Table 1.

[30] Ibid., App. 6, Table 46, and para. 89.

[31] Ibid., paras. 37–43. In 23 cases the conviction was quashed by the Court of Appeal.

[32] In the years 1900–49, while there were 7,454 murders known to the police, the total number of convictions was 1,246; of these thirty-four youths under eighteen and two pregnant women were not sentenced to death. Of the 1,210 death sentences passed, 23 were quashed by the Court of Appeal. The number of convictions, 1,246, looks small in comparison with the total numbers of murders known to the police, 7,454 of which 2,001 were babies under one year. But during these same 50 years 1,674 suspects committed suicide and 1,226 persons were found either guilty but insane and acquitted or unfit to plead. Ibid., App. 3, Table 1. In terms of annual average, for every 149 murders known to the police (or 109, if babies under one year are excluded) there are 25 convictions, 33 suicides by suspects and 25 persons are found guilty but insane or unfit to plead. During the entire period 1900–49 3,130 persons were charged with murder, an annual average of 63.

The scale of criminal homicide in the United States is very different from the humble English figures. According to the note on the classification of offences included in all issues of the *Uniform Crime Reports*, 'murder and non-negligent manslaughter' is said to comprise all wilful felonious homicides as distinguished from deaths caused by negligence. Accordingly, when faced with a corpse the police must, before classifying the case under this head, exclude the possibilities of negligent homicide, suicide, accidental death, and justifiable homicide (defined in the *Reports* as the killing of a felon by a police officer in line of duty or the killing of a hold-up man by a private citizen). Each year the *Uniform Crime Reports* then purport to *estimate* the total number of murders and non-negligent manslaughters in the whole country on the basis of figures actually reported from police authorities now covering approximately four-fifths of the total population. The *Reports* claim that these estimates are 'conservative' indications of the 'nationwide major crime problem'.[33]

To make some comparisons with England and Wales I have chosen the ten-year period from 1945 to 1954, during which the estimated total of offences under the head of murder and non-negligent manslaughter in the United States was 72,679: this gives a yearly average (to the nearest unit) of 7,268.[34]

[33] See, e.g. 26 *Uniform Crime Reports*, 72 (1956).

[34] The annual estimated total appears each year as the first table in the second semi-annual bulletin of the *Uniform Crime Reports*. The separate figures for these years together with the estimated population for the United States taken from *Bureau of Census, Dept of Commerce, Statistical Abstract of the United States* (1955) p. 13, Table 8 are as follows:

Year	*Murder and Non-Negligent Manslaughter*	*Total Population Residing in U.S.A. (in thousands)*
1945	6,847	132,481
1946	8,442	140,054
1947	7,760	143,446
1948	7,620	146,093
1949	6,990	148,665
1950	7,020	151,234
1951	6,820	153,384
1952	7,210	155,755
1953	7,120	158,306
1954	6,850	161,195

These figures, expressed (as the English figures are) as rates per *million* of population, vary during these ten years between 60 and 40 per million. This is between fifteen and ten times the rate in England and Wales for the fifty-year period of 1900–49. It should be remembered, however, that certain types of homicide included under the American classification of 'murder and non-negligent manslaughter' are excluded from the British figures, which are for murder alone.[35] The chief, if not the only, cases of non-negligent manslaughter are those where a plea of provocation might be successfully maintained, but unfortunately it is not possible to estimate separate figures for 'murders' and 'non-negligent manslaughters'. Perhaps, especially in the southern states, the number of instances of provoked homicides which would be treated as manslaughter by the courts might be considerable.

In many parts of the country, particularly in the southern states, the rates of murder and non-negligent homicide are very much higher than the average for the whole country quoted above and amount to as much as fifty times the English murder rates; in other parts of the country the rates are much lower than the national average. Thus if we take the urban centres in Georgia for the three years 1950-2, the respective rates per million of population are 177.9, 182.3, and 206.7, while the rate in New Hampshire urban centres is 7.4, 4.1, and 19.5 per million. For the same three years the rates for urban centres, covering a population of about nineteen million, in the five states of Illinois, Michigan, Indiana, Missouri, and Wisconsin were respectively 42.1, 48.3, and 45.8 per million.

According to 25 Uniform Crime Reports 69 (1954), in the 20 years from 1935–54, 146,869 persons (an annual average of 7,344) were wilfully and unlawfully slain in the United States. The highest year in these twenty was 1946 with 8,442 wilful killings. For each one million persons there were 61 murders in 1935 and 42 in 1954.

[35] The English criminal statistics give separate figures for murders known to the police, but do not distinguish between negligent and non-negligent manslaughter. The combined annual figures for both murder and manslaughter (negligent as well as non-negligent) in England and in Wales during 1900–49 were about 8 per million inhabitants, while in the same period in about 2,200 cities in the United States the number of murders and non-negligent manslaughters was about 56 per million. See *Royal Commission Report*, App. 6, paras. 12, 24, 88.

These figures show how different the magnitude of the murder problem in the United States is from the problem in England, and it would be possible to present this difference in many dramatic ways. In Chicago alone, for example, the *number* of murders and non-negligent homicides in each of the three years 1950–2 (257, 249, and 289) was nearly double the number of murders in the whole of England and Wales (139, 132, and 146).[36]

During the ten-year period between 1945 and 1954, 775 persons were executed for murder by the civil authorities, an average of seventy-seven per year. This, compared with the average annual estimated number of murders and non-negligent manslaughters for those years (7,268), gives a ratio of less than 1:100;[37] in England the ratio of executions to murders known to the police for the fifty years 1900–49 was about 1:12.

Though we lack satisfactory figures, most authorities share the view that the number of murders committed by professional criminals is far greater in the United States than it is in England; a far greater proportion of murder in England is due to jealousy, drink, quarrel, lust, and even irritation than in the United States. The Royal Commission estimated that for the period 1900–49 20 per cent of convicted murderers in England, at the most, were professional criminals.[38] In England insanity, defined even by the stringent legal criteria used (until 1957) for assessing criminal responsibility, plays a very great part: of the total of 3,129 persons committed to trial for murder during the fifty years 1900–49, 428 were held unfit to plead and 798 adjudged guilty but insane under the M'Naghten rules. The combined figures for these two

[36] All American figures in this paragraph are from 21 *Uniform Crime Reports*, 90, 92, 95 (1950); 22 ibid. at 87, 89, 93 (1951); 23 ibid. at 93, 95, 99 (1952). For England and Wales see *Royal Commission Report*, App. 3, Table 1.

[37] See *Bureau of Census, Dept of Commerce, Statistical Abstract of the United States* (1955) p. 154, Table 188.

[38] See *Royal Commission Report*, App. 6, Table 2 and paras. 12, 13. In 1930 Judge Marcus Kavanagh told the Select Committee on Capital Punishment that 'the larger number of people who are killed in the United States are killed by criminals.' Ibid. para. 12.

categories of insanity (1,226) were slightly greater than the total of those convicted and sentenced to death for murder (1,210) during this period.[39] The relevant figures for the United States are apparently not available.

4

Let us now turn to the principles to which men appeal when they argue for or against the death penalty. In any public discussion of this subject the question that is likely to be the central one is 'What is the character and weight of the evidence that the death penalty is required for the protection of society? What is the evidence that it has a uniquely deterrent force compared with the alternative of imprisonment?'[40] Later we shall examine what evidence there is to answer this question, but first we should consider what is implied if this question is treated—as undoubtedly most ordinary men now do treat it—as the root of the matter, as the fundamental question in considering whether the death penalty should be abolished or retained.

To treat this question as the root of the matter is implicitly to adopt what is called, I think unhappily, a theory of

[39] *Royal Commission Report*, App. 3, Table 1. Respite was granted an additional forty-seven persons, after sentence of death, by the Home Secretary on grounds of insanity. Ibid., at p. 301.

[40] A careful examination of the English Parliamentary debates confirms this, although it is certainly not apparent at first sight. Thus the Archbishop of York, in the Lords debate in July 1956, insisted 'on the moral necessity of retribution within our penal code'. The Lord Chancellor agreed with the view elsewhere expressed by Lord Justice Denning that 'the ultimate justification of any punishment is not that it is a deterrent, but that it is the emphatic denunciation by the community of a crime'. 198 *H.L. Deb.* 576 (1956). But the Lord Chancellor also said that 'the real crux' of the question at issue is whether capital punishment is a uniquely effective deterrent; ibid, at 577. The Archbishop stated that 'the question of deterrence comes to the head as a vitally important matter'; ibid., at 597. See also Lord Salisbury (a retentionist): 'For me, as for many others, it is on the deterrent value of capital punishment that the whole balance of the argument must turn'; Ibid., at 820. For an illuminating philosophical analysis of the arguments in this debate see Gallie, 'The Lords Debate on Hanging, July 1956: Interpretation and Comment,' *Philosophy*, Vol. 32 (Apr. 1957), p. 132.

punishment. I say 'unhappily' because theories of punishment are not theories in any normal sense. They are not, as scientific theories are, assertions or contentions as to what is or what is not the case; the atomic theory or the kinetic theory of gases is a theory of this sort. On the contrary, those major positions concerning punishment which are called deterrent or retributive or reformative 'theories' of punishment are moral *claims* as to what justifies the practice of punishment—claims as to why, morally, it *should* or *may* be used. Accordingly, if it is held that the central question concerning the death penalty is whether or not it is needed to protect society from harm, then, although *this* question is itself a question of fact, the moral claim (or 'theory' of punishment) implied is the 'utilitarian' position that what justifies the practice of punishment is its propensity to protect society from harm. Let us call this implicit moral claim 'the utilitarian position'.

There are indeed ways of defending and criticizing the death penalty which are quite independent of the utilitarian position and of the questions of fact which the utilitarian will consider as crucial. These are perhaps more commonly expressed in England than in America. For some people the death penalty is ruled out entirely as something absolutely evil which, like torture, should never be used however many lives it might save. Those who take this view find that they are sometimes met by the counter-assertion that the death penalty is something which morality actually demands, a uniquely appropriate means of retribution or 'reprobation' for the worst of crimes, even if its use adds nothing to the protection of human life. Here we have two sharply opposed yet similar attitudes: for the one the death penalty is morally excluded; for the other it is a moral necessity: but both alike are independent of any question of fact or evidence as to what the use of the death penalty does by way of furthering the protection of society. Argument in support of views as absolute as these can consist only of an invitation, on the one hand, to consider in detail the execution of a human being, and on the other hand, to consider in detail some awful murder, and then to await the emergence either of a conviction that the death penalty must never be used or, alternatively, that it must never be completely abandoned.

It is important to realize that what differentiates the utilitarian position from these absolute attitudes is not that the latter adopt a specific moral attitude while the utilitarian position confines itself to 'the facts'. The utilitarian position, which treats the welfare of society as the justification of punishment, is also a moral claim just as these absolute positions are; what differentiates them is that the utilitarian position commits one, as the absolute positions do not, to a factual inquiry as to the effects upon society of the use of the death penalty.

Is the utilitarian position coherent? Is it possible to hold it without paradox or without commitment to consequences against which most ordinary people's moral sense would rebel? Or when the consequences of the utilitarian position are exposed do men feel compelled by other moral principles which they hold at least as firmly to abandon or qualify the utilitarian position? What are these other moral principles? Do they imply the tacit admission that something going under the ambiguous name of 'retribution' or 'reprobation' requires attention in any acceptable 'theory' of punishment? If so, to what specific aspect of punishment are these notions relevant? I think that most of the puzzles about the principles of punishment which trouble ordinary men can be reduced to these questions.

Let us consider some of the claims that are urged against the utilitarian position. They are often obscurely presented and I shall try to put them clearly.

The first is this. It is often said that men punish and always have punished for a vast number of different reasons. They have punished to secure obedience to laws, to gratify feelings of revenge, to satisfy a public demand for severe reprisals for outrageous crimes, because they believed a deity demands punishment, to match with suffering the moral evil inherent in the perpetration of a crime, or simply out of respect for tradition. If there are these many reasons, why should we select the protection of society from harm and give this primacy as the 'basis' of punishment? Surely it is only one reason among many which stand on an equal level in so far as a claim upon our attention is concerned.

Here plainly we must distinguish two questions commonly confused. They are, first 'Why do men in fact punish?' This is a question of fact to which there may be many different answers such as those exemplified above. The second question, to be carefully distinguished from the first, is 'What justifies men in punishing? Why is it morally good or morally permissible for them to punish?' It is clear that no demonstration that in fact men have punished or do punish for certain reasons can amount *per se* to a justification for this practice unless we subscribe to what is itself a most implausible moral position, namely, that whatever is generally done is justified or morally right. Short of this, if we think that punishment is *justified* because, for example, it satisfies a public demand or because it meets the evil of misconduct with suffering, we must add to our statement of fact that men in fact do punish for such reasons, the further *moral* claim that it is good or at least morally permissible to punish for such reasons.

When this simple point is made clear and the two questions 'Why do men punish?' and 'What justifies punishment?' are forced apart, very often the objector to the utilitarian position will turn out to be a utilitarian of a wider and perhaps more imaginative sort. He will perhaps say that what justifies punishment is that it satisfies a popular demand (perhaps even for revenge) and explain that it is good that it satisfies this demand because if it did not there would be disorder in society, disrespect for the law, or even lynching. Such a point of view, of course, raises disputable qustions of fact as to the extent to which satisfaction of popular demand is important in the ways indicated. None the less, this objection itself turns out to be a utilitarian position, emphasizing that the good to be secured by punishment must not be narrowly conceived as simply protecting society from the harm represented by the particular type of crime punished, but also as a protection from a wider set of injuries to society.

Very often, however, because the question of fact and the question of justification are not thus distinguished, the fact, or the alleged fact, of a public demand for punishment (or a particular kind of punishment) is cited as if it were *per se* a justification; or, at least, the precise moral principle which treats such a demand (or some element of it) as a justification for punishment is never clearly stated or exposed for criticism.

But the objector who criticizes the utilitarian position by reminding us of the diversity of reasons for which men punish may not always turn out to be a wider utilitarian in the way I have suggested. Sometimes the objector will take his stand on absolutes and claim that meeting the moral evil of misconduct with suffering is, as Kant urged, good *per se*, so that, even on the last day of society, the murderer not only may but must be executed. But before we say that no argument is possible between the utilitarian and objectors of this sort, it is necessary to inquire whether the objector would hold his position unless he also believed that punishment was necessary to protect society from harm. Would he really rely on his absolutist position to justify going *beyond* the limits of what the utilitarian would admit by way of punishment, and inflict a punishment more severe than one required on utilitarian principles?[41] Sometimes the answer is 'yes' and then we are left to a clash of fundamental moral claims in which the absolutist must simply expose for inspection and acceptance his claim that there is somehow some intrinsic total good in meeting the moral evil of misconduct with suffering; this, he must say, is something morally 'called for' independently of its place in a social mechanism designed for the protection of society or other beneficial effects.

Consider now a more fundamental objection, to the utilitarian position which is implied in holding the central question in relation to the death penalty to be the question 'What is the evidence that it is needed to protect society?' 'Surely', says the objector, 'the protection of society cannot be your justification, for if it were, why should you stop where you not only do stop, but think you ought to stop, in using punishment as an

[41] It seems that the Archbishop of York and the Lord Chancellor, while insisting that the primary purpose of punishment is retribution (see *supra*, p. 71, n. 40), would have said 'no' at this point.

instrument for the protection of society? Why not employ torture if that would effectively stop, e.g., parking offences; or at any rate why not employ a punishment immeasurably more severe than we normally contemplate for this type of offence? Why not, if in a particular case it were necessary and possible in order to protect society, put up an innocent man, fake his guilt and execute him? What moral, as distinguished from practical, objection could there be on utilitarian grounds to the staged trials or to the shooting of the innocent *pour encourager les autres*? Does not the common insistence that punishment be applied only to one who has in fact broken the law and, in the case of serious crime, done so with *mens rea* show that you are guided by considerations quite different from utilitarian principles?'

Here we must go carefully; for in this type of objection, which certainly troubles the plain man's utilitarianism, many different issues are involved. Clearly it is part of a *sane* utilitarianism that no punishment must cause more misery than the offence unchecked; and it might well be that the misery caused to the victim and his friends by torture or other very severe punishments would be worse than any misery caused by a minor offence for which it was used as a punishment. No doubt also a consistent utilitarian answer could be given to the other objections. The state of general alarm and terror which might arise in society if it were known that the innocent were likely to be seized and subjected to the pains of punishment in order to serve the needs of society might be worse than any advance in security or social welfare brought about by these means could outweigh. Furthermore, administrative and judicial officers might refuse to give effect to the use of 'punishment' in such circumstances and would hence 'nullify' it.

Yet though such answers *can* be made they do not seem to account for the character of the normal unwillingness to 'punish' those who have not broken the law at all, nor for the moral objection to strict liability which permits the punishment of those who act without *mens rea*. We cannot be so easily rid of the argument that some elements other than those which even the broadest utilitarian admits are involved. Bentham himself confronted the doctrine of *mens rea* and asked why we do and should excuse from criminal responsibility persons who have committed a crime owing to their mental condition, either temporary (mistake, accident, duress, &c.) or relatively enduring (insanity, infancy). He thought that it was enough to say that the *threat* of punishment would here be socially useless. It could not deter such persons or other people like them and hence on plain utilitarian grounds, which enjoin us not to cause useless suffering, they should be excused from punishment. He even went so far as to say that this is all that could be *meant* by the restriction of punishment to those who have a 'vitious will' (as Blackstone had termed *mens rea*).[42] But there is a *non sequitur* in Bentham's argument. He claims to show that *punishment* of such persons as we excuse on such grounds would be wrong because it would be socially useless ('inefficacious'), whereas he only shows that the *threat* of punishment would be ineffective so far as such persons are concerned. Their *actual* punishment might well be 'useful' in Benthamite terms because, if we admit such excuses, crime may be committed in the hope (surely sometimes realized) that a false plea of mistake, accident, or mental aberration would succeed.

Apart from this inconclusive argument, this unqualified utilitarianism does not reproduce the real moral objection that most thinking people have to the application of the pains of punishment to the innocent or to those who, by reason of their mental condition, are thought unable to comply with the law's demands. This moral objection normally would be couched as the insistence that it is *unjust*, or *unfair*, to take someone who has not broken the law, or who was unable to comply with it, and use him as a mere instrument to protect society and increase its welfare. Such an objection in the name of *fairness* or *justice* to individuals would still remain even if we were certain that in the case of the 'punishment' of one who had not broken the law the fact of his innocence would not get

[42] See Bentham, *An Introduction to The Principles of Morals and Legislation*, Chap. XIII, 'Cases Unmeet for Punishment'.

out or would not cause great alarm if it did. Similarly, even if it were shown that the admission as an excuse of types of insanity or other defences in fact led, owing to successful fake pleas, to utilitarian 'losses', i.e. to a greater prevalence of crime than would be the case if the system allowed for no such excuse, we would still be morally reluctant to allow punishment in such cases. To our tolerance of such a system there would indeed be some limit; but even if we were convinced that the social danger of the evasion of punishment through false pleas were overwhelming, and were forced to extend the area of strict liability, we would *wittingly* be choosing between two distinct principles: the utilitarian principle which justifies punishment by its propensity to protect society from harm, and a principle of justice which requires us to confine punishment to those who have broken the law and had at least some minimum capacity to comply with it.[43] Hence it still remains for the utilitarian to give some coherent

[43] The distinction between the efficacy of (1) the *threat* of punishment and (2) the *actual* punishment should be remembered when modern restatements of the Benthamite rationale of punishment are considered. This is true of the best of such modern restatements such as Michael and Wechsler's well known 'A Rationale of the Law of Homicide' 37 *C. L. R.* (1937) pp. 701, 1261, and *Criminal Law and its Administration* (1940). See also Wechsler, Book Review, 37 *C. L. R.* (1937) p. 687.

On the whole these authors consistently identify the criminally 'responsible' with those whom the threat of punishment *could* deter or as they sometimes phrase it are 'capable of choosing to avoid the act in order to avoid punishment'. This class, of course, is not necessarily the same as those whose *punishment* might benefit society (whether by deterring them, or others by example or both) and may (for the reasons stated above) well be a narrower class. But certainly to confine punishment to those who *could* be deterred by the threat (though *in fact* they were not deterred) accords with the common conviction that it would be unjust or unfair to punish those who could not be influenced by the threat of punishment; whereas a policy of punishing all those whose *actual* punishment might be socially useful would not accord with *this* conviction. On the other hand, it is by no means clear that a theory of punishment which thus restricts punishment to those who could be but actually were not deterred by the threat of punishment evades all the difficulties of 'free will' as these authors suggest (e.g. 37 *C. L. R.,* at p. 690). For *prima facie* at any rate the statement that some one 'could have been' deterred from a crime which he in fact committed means that in *the actual circumstances* in which he committed the crime he could have acted otherwise; and surely the classical 'problem' of free will is just whether we ever have a right to make any such statement. But as these authors claim, and as I argue above, the restriction of punishment to those who have committed crimes voluntarily (i.e., not under the usual excusing conditions of mistake, accident,

reason why he should object to the use of punishment in the way suggested, and the question for him is whether he can do this without adopting the theory that the fundamental justification of punishment is not the protection of society but the return of suffering for the moral evil done. Must he, in order to make sense of his refusal to punish those who have not broken the law, adopt the notions of retribution, reprobation, expiation, or atonement? Is it true, as his opponent claims, that the only reason we can have for restricting the use of punishment in the way suggested is the moral conviction that what justifies punishment is a return of suffering for moral guilt.

There is a coherent answer which a cautious utilitarian can make to this objection without admitting notions of retribution, unless 'retribution' means merely that punishment must be confined to those who have broken the law and could have helped this; but like all objections which are recurrent in the history of an idea this one shows something important. It shows that the utilitarian position, to be plausible, must be

insanity, or the like) can be explained perfectly well without resort to retributive 'theories' of punishment.

The authors carefully consider the possibility that loyalty to the principle that only the deterrable be punished might lead to the admission of certain excuses, e.g. the 'irresistible impulse' test of insanity, and this might weaken the deterrent effect of the law upon those who could be deterred but hope successfully to evade punishment by simulating this excuse. They urge that the danger of this weakening the 'net deterrent efficacy' of the law is not 'decisive' in favour of rejecting this excuse, since its *rejection* might also lead to socially undesirable results in the form of e.g. nullification of the system and 'public excitement' which the execution of clearly undeterrable persons might produce. See Ibid., at 752–7.

But it is important here to emphasize (as these authors do not) that there are moral objections (at least as firm as any utilitarian principles) to punishing persons who are clearly undeterrable (incapable of effective choice) and these objections are *not* merely subordinate aspects of the social desirability of avoiding public excitement and nullification of the system, etc. Indeed the reason why we should expect 'public excitement' or the nullification of a system which permitted 'the undeterrables' to be executed is precisely because it is widely considered (independently of social welfare) *unfair* or *unjust* to punish them. A theory of punishment which disregarded these moral convictions or viewed them simply as factors, frustration of which made for socially undesirable excitement is a different kind of theory from one which *out of deference to those convictions themselves* restricts punishment to those who are deterrable or capable of acting so as to avoid punishment.

regarded as a claim to the *outer* limits of punishment; as fixing a *maximum* beyond which punishment is not justified. The utilitarian position, in however sophisticated a version, cannot plausibly be regarded as something which we can use in an unqualified fashion. There are many different ways in which we think it morally incumbent on us to *qualify* or *limit* the pursuit of the utilitarian goal by the methods of punishment. Some punishments are ruled out as too barbarous or horrible to be used whatever their social utility; we also limit punishments in order to maintain a scale for different offences which reflects, albeit very roughly, the distinction felt between the moral gravity of these offences. Thus we make some approximation to the ideal of justice of treating morally like cases alike and morally different ones differently.

Much more important than these is the qualification which civilized moral thought places on the pursuit of the utilitarian goal by the demand that punishment should not be applied to the innocent; indeed, so insistent is this demand that no system of rules which generally provided for the application of punishment to the innocent would normally be called a system of punishment. But the moral basis of this claim that such a limit must be imposed on the pursuit of the utilitarian goal need not be, and in most ordinary persons' minds is not, a recognition that the fundamental justification of punishment is other than the pursuit of the utilitarian goal. To see this point clearly we again must distinguish two very different types of questions. The first is 'What justifies the general practice of punishment?', and the utilitarian answer that the justification lies in the need to protect society from harm is, by itself, an adequate answer to this question. There is, however, a further question: 'What justifies us in applying the system of punishment to a particular individual?' Something more is involved in this question, for a necessary condition of the just *application* of punishment to a particular individual includes the requirement that he has broken the law. There are many ways of presenting this distinction: it is the distinction between the justification of punishment as a practice and the liability to punishment, or the distinction between the general question 'What justifies us in maintaining laws by the practice of punishment?' and the particular question 'Who may be punished?' It is important to see that, while the conviction that something more is required when we come to the particular question than sufficed for the general question is sound, it is not a recognition of an alternative basis or justification for the general practice of punishment. For the stipulation that punishment should not be applied except to an individual who has broken the law, may be made not to secure that moral evil should meet its return in punishment, but to protect the individual from society. It may be the recognition of the claim of the individual that he should not be sacrificed for the welfare of society unless he has broken its law; his breach of the law is, as it were, a condition or a licence showing us when there is liability to punishment. It is not an alternative basis for the system and could not (as a retributive or reprobative theory could) justify our using penalties more severe than would be required on utilitarian grounds. No doubt this recognition of the individual's claim not to be sacrificed to society except where he has broken laws is not itself absolute. Given enough misery to be avoided by the sacrifice of an innocent person, there may be situations in which it might be thought morally permissible to take this step. But, again, if we took the step, we would have to face a clash between two principles. We would then sacrifice the principle of fairness designed to protect the individual from society to the principle that an overwhelming advantage to society should be secured at any cost; but a clash between two principles is different from the simple application of a single utilitarian principle that anything which benefits society is permissible.

Of course, the distinction just emphasized between (1) a utilitarian justification of punishment qualified by recognition of the innocent individual's claim not to be sacrificed to society, and (2) a frankly retributive 'theory' in which punishment is justified simply as a return for moral evil, must become meaningless if crime is regarded (as it is in some contemporary thought) always and only as a disease, with the corollary that it should be treated always and only with preventatives and cures. From this point of view a utilitarianism qualified by the principle that it is just only to inflict punishment on those who have voluntarily broken the law is as absurd as a theory that says, as an extreme retributivist theory does, that the justification for punishment is simply to match the past evil of misconduct with the pain of suffering. Indeed, the notion of fairness or justice would be almost senseless in the context of this outlook, for the contention that it is fair or just to punish those who have broken the law must be absurd if the crime is merely a manifestation of a disease. This point of view, which in effect would replace the notion of punishment with the idea of social hygiene, may rest either on philosophical determinism, or on the conviction that in no case where a crime has been committed are there any adequate empirical grounds for the belief that the criminal could have done other than he did. These viewpoints are not, of course, to be dismissed lightly, but they can be discussed only if we are prepared to drop the whole range of concepts involved in the institution of punishment. Necessarily these considerations are outside the confines of this discussion of a specific form of punishment.

If we look back on this discussion it appears that the utilitarianism of the plain man, if it is to be tenable, must be qualified in the face of the question: 'Why not punish the innocent if in a given case it promotes the welfare of society?' The qualification to be made is the admission that the individual has a valid claim not to be made the instrument of society's welfare unless he has broken its laws; but to recognize this qualification of utilitarianism is not to recognize a different basis or justification for the practice of punishment.

There has emerged from this consideration of punishment, therefore, a need to distinguish between two pairs of distinct questions. The first set brings out the difference between asking 'Why do men punish in fact?' and 'What justifies them in doing so?' The second brings out the difference between asking 'What justifies in general the practice of punishment?' and 'What more is required, given that there is this general justification for the practice, in order to justify its use in any particular case?'

5

Let us now return to the central question: 'What is the weight and character of the evidence that the death penalty is required for the protection of society?' Here there are two main approaches and both of them are strewn with pitfalls. One of them is through statistics, the other through what has been termed a 'common-sense' conception of the strength of the fear of death as a motive in human conduct.

Statistics have now been collected and surveyed in a more thorough fashion than ever before. Yet the Report of the Royal Commission, after considering the expert scrutiny of the figures available in Europe, the Commonwealth, and the United States, reached only a negative, though still an important, conclusion. This was the finding that there is no clear evidence in any of the figures that the abolition of the death penalty has ever led to an increase in the rate of homicide or that its restoration has ever led to a fall.[44] Important as this is, it is of equal importance to appreciate that this investigation also showed how little we know, and perhaps can ever know, about the effect of the penalty on social life.

There are three cardinal points.

(1) Comparisons between countries which retain the death penalty and countries which have abolished it are practically useless. The rate in death penalty England is lower than the rate in the abolition Scandinavian countries;[45] the rate in abolition Wisconsin is higher than death

[44] *Royal Commission Report*, para. 65.

[45] The rates per million of population of murders in England and Wales and 'intentional homicide' in Norway for three decades in 1910–1940 were as follows:

England and Wales	4·1 (1910–1919)	3·9 (1920–1929)	3·3 (1930–1939)
Norway	5·4 (1911–1920)	4·9 (1921–1930)	5·0 (1931–1940)

The Norwegian figures do not include babies, which constituted 28.5% of the English figures. See *Royal Commission Report* App. 3 Table 1, App. 6 s. 89, Table 46.

penalty England but lower than many death penalty states in the United States. Obviously differences in population, in social, economic, and psychological conditions may render fallible any inference from the experiences of one country as to what may be expected from the death penalty or its abolition in any other.

(ii) The only rational use of the figures is to compare the statistical history of one country before and after abolition, or before and after the introduction or reintroduction of the death penalty, and to ask whether there are any changes in the murder rate correlated with these changes in the penalty. But there are many pitfalls here which reduce the utility of the available statistics. The foremost of them in importance are these: (*a*) In many countries formal abolition came only after a long period of gradual desuetude. In Norway, for example, the last execution was in 1876 but formal abolition came only in 1905, and such has been the pattern of many abolition countries in Europe and the Commonwealth. Where this 'gradualness' obtains it is difficult to estimate when the death penalty ceased in practice to be a serious threat. (*b*) In any case, even if the point at which the death penalty either ceased to be or became a serious threat could be precisely marked, its operation on the murder rate is likely to be a long term effect: 'There is unlikely to be in any civilised country a string of would-be murderers straining at the leash waiting only for the death penalty to be removed to commit murder: or vice versa. The effect is likely to be cumulative.'[46]

(iii) The best and most impressive types of evidence come from cases where one of a bloc of several neighbouring states of similar population and similar social and economic conditions has abolished or introduced the death penalty for murder while the others have not changed it. Nebraska and North and South Dakota are examples of such a bloc, and the rise and fall of the murder rate in

[46] Gold, 'Should the Death Penalty be Abolished?' Letter in *Listener* (9 Feb. 1956), p. 217.

these three states was much the same during the period 1930–48, although South Dakota reintroduced the death penalty in 1939 after previous abolition, Nebraska retained capital punishment (but made use of it only twice in this period), and North Dakota was an abolition state. Such comparisons between fairly homogeneous states suggest that the murder rates in such states are conditioned by factors operating independently of the death penalty.[47] There is, however, too little of such evidence to justify a positive inference.

In fact, perhaps the most important lesson from a dispassionate survey of the statistics is the need to distinguish between the two following propositions. (1) There is no evidence from the statistics that the death penalty is a superior deterrent to imprisonment. (2) There is evidence that the death penalty is not a superior deterrent to imprisonment. The Commission's conclusion is strictly confined to the first of these propositions, though many advocates of abolition speak as if the second were a warranted conclusion from the figures. That this is not so may be dramatically illustrated from the following facts. In the thirty years from 1910 to 1939 the ten-year average murder rate in England fell from 4.1 to 3.3 per million. Yet if the death penalty had been abolished at the beginning of this period (1910), and if this had resulted in 100 more murders than there actually were during this period, there would still have been a substantial decrease (from 4.1 to 3.5 per million) in the murder rate following this abolition.[48] We would have said 'in this case abolition was not followed by an increase but by a decrease in the murder rate' and have been tempted to treat this as evidence that there was no connexion between the rate of crime and the form of penalty. This serves to show the importance of presenting our conclusion in the negative form that there is no evidence from the figures in favour of capital punishment.

If we turn from the statistical evidence to the other 'evidence', the latter really amounts simply to the alleged truism that

[47] *Royal Commission Report*, para. 64, App. paras. 51–54, Tables 24–28.

[48] Gold, *supra*, n. 46, at p. 217.

men fear death more than any other penalty, and that therefore it *must* be a stronger deterrent than imprisonment. No one has proclaimed his faith in this proposition more strongly than the great Victorian judge and historian of the Criminal Law, James Fitzjames Stephen. He said:

> No other punishment deters men so effectually from committing crimes as the punishment of death. This is one of those propositions which it is difficult to prove, simply because they are in themselves more obvious than any proof can make them. It is possible to display ingenuity in arguing against it, but that is all. The whole experience of mankind is in the other direction. The threat of *instant* death is the one to which resort has always been made when there was an absolute necessity for producing some result. . . . No one goes to *certain inevitable* death except by compulsion. Put the matter the other way. Was there ever yet a criminal who, when sentenced to death and *brought out to die,* would refuse the offer of a commutation of his sentence for the severest secondary punishment? Surely not. Why is this? It can only be because 'All that a man has will he give for his life'. In any secondary punishment, however terrible, there is hope; but death is death; its terrors cannot be described more forcibly.[49]

This estimate of the paramount place in human motivation of the fear of death reads impressively, but surely it contains a *suggestio falsi* and once this is detected its cogency as an argument in favour of the death penalty for murder vanishes. For there is really no parallel between the situation of a convicted murderer offered the alternative of life imprisonment in the shadow of the gallows, and the situation of the murderer contemplating his crime. The certainty of death is one thing; perhaps for normal people nothing can be compared with it. But the existence of the death penalty does not mean for the murderer *certainty* of death *now*; it means a not very high probability of death in the future. And futurity and uncertainty, the hope of an escape, rational or irrational, vastly diminishes the difference between death and imprisonment as deterrents, and may diminish it to vanishing point. And the hope of escape is not so very irrational even in America or in the best policed states. In England, if we compare the number of murders known to the police with the number of convictions and executions, the chance of conviction appears to be one in

[49] Stephen, 'Capital Punishments', *Fraser's Magazine,* Vol. 69 (1864), p. 753 (my italics), quoted in the *Royal Commission Report,* para. 57.

six, and the chance of execution one in twelve. If, however, we assume that the very large number of suspects who commit suicide would have been caught anyway, the chance of conviction increases to one in three, and of execution to one in six. It would, of course, be ridiculous to think that these figures are appreciated by potential murderers, but they do serve to show that the way in which a convicted murderer may view the immediate prospect of the gallows after he has been caught must be a poor guide to the effect of this prospect upon him when he is contemplating committing his crime.

But there is a more important reason why this insistence on the unique status of the fear of death as a motive helps us very little here. In all countries murder is committed to a very large extent either by persons who, though sane, do not in fact count the cost, or are so mentally deranged that they cannot count it. In all countries the proportion of 'insane' murderers is very high, and in England and Wales in the fifty years from 1900 to 1949 the numbers of those who were charged with murder and found insane, by very stringent tests, exceeded the number of persons who were sentenced to death. In England, moreover, for every four murders known to the police approximately one suspect commits suicide, and it is likely that many of these suicides had made up their mind to die before they committed the crime.

For these reasons, many would not attach even as much weight as did the Royal Commission to what they term the common-sense argument from human nature. The Commission said:

> . . . *Prima facie* the penalty of death is likely to have a stronger effect as a deterrent to normal human beings than any other form of punishment, and there is some evidence (though no convincing statistical evidence) that this is in fact so. But this effect does not operate universally or uniformly, and there are many offenders on whom it is limited and may often be negligible. It is accordingly important to view this question in just perspective and not to base a penal policy in relation to murder on exaggerated estimates of the uniquely deterrent force of the death penalty.[50]

[50] *Royal Commission Report,* para. 68.

Certainly if as much weight as this is attached to the 'common-sense' argument it is necessary to remember other aspects of the death penalty. One day, indeed, the still young sciences of psychology and sociology may confirm the speculation that the fear of death has the potency thus claimed for it, or perhaps that the death penalty has had some unique influence in building up and maintaining our moral attitude to murder. But we certainly cannot take this to be established, and those who base their advocacy of the death penalty on this rough 'common-sense' psychology must seriously consider psychological theories that run in the other direction. For at present, theories that the death penalty may operate as a stimulant to murder, consciously or unconsciously, have some evidence behind them. The use of the death penalty by the state may lower, not sustain, the respect for life. Very large numbers of murderers are mentally unstable, and in them at least the bare thought of execution, the drama and the notoriety of a trial, the gladiatorial element of the murderer fighting for his life, may operate as an attractive force, not as a repulsive one. There are actual cases of murder so motivated, and the psychological theories which draw upon them must be weighed against the theory that the use of the death penalty creates or sustains our inhibition against murder.

6

What, then, is the final upshot? My purpose has been to lay bare the known facts and relevant principles and not, of course, to press upon the reader the inferences which I would draw in considering the question of abolition or retention in England. I shall, however, add this very simple final consideration. If we adopt the kind of qualified utilitarian attitude toward punishment which appears to me to accord (as an unqualified utilitarianism does not) with the moral convictions which most of us share, then it is vital to consider where the onus of proof lies in this matter of the death penalty.

Is it upon those who object to the death penalty to show positive evidence that it is socially useless, if not harmful? Or is it upon those who would retain it to show that it is socially beneficial? Three main factors made the death penalty and the mode of its use in England appear a *prima facie* evil and therefore only to be retained if there was some positive evidence that it was required in order to minimize murder, or because it served some other valuable purpose which other punishments could not serve. These factors were: (1) *prima facie* the taking of a life, even by the State, with its attendant suffering not only for the criminal but for many others, is an evil to be endured only for the sake of some good; (2) the death penalty is irrevocable and the risk of an innocent person being executed is never negligible;[51] and (3) the use of the death penalty in England was possible only at the cost of constant intervention by the Executive after the courts had tried and sentenced the prisoner to death. Of course, it is possible that sincere and thoughtful men may differ in their moral estimation of these three factors. But the first two of these factors are as applicable in the United States as in England, and there is some analogy to the third factor in the possibility, inescapable in the United States, and sometimes realized, of long periods intervening between the sentence of death and its execution.

[51] For Parliamentary discussion of the possibility of mistake see 548 *H.C. Deb.* 2540–8, 2557–9, 2583, 2597–8 (1956). Mr Chuter Ede who, as Home Secretary, had himself refused a reprieve in the well-known case of Timothy Evans, stated subsequently that he no longer thought that Evans was guilty. In arguing in the debate in February 1956, in favour of abolition, he claimed that if before Evans's execution evidence had been available which came to light after the execution, public opinion would not have allowed the execution to have taken place. This case and Mr. Ede's statement (since he had in 1948 as Home Secretary urged retention of the death penalty) must have weighed with many who voted for suspension. 548 *H.C. Deb.* 2558–9, 2583 (1956).

IV

ACTS OF WILL AND RESPONSIBILITY

The General Doctrine

IN this lecture I propose to air some doubts which I have long felt about a doctrine, concerning criminal responsibility, which has descended from the philosophy of conduct of the eighteenth century, through Austin, to modern English writers on the criminal law. This is the doctrine that, besides the elements of knowledge of circumstances and foresight of consequences, in terms of which many writers define *mens rea,* there is another 'mental' or at least psychological element which is required for responsibility: the accused's 'conduct' (including his omissions where these are criminally punishable) must, so it is said, be voluntary and not involuntary. This element in responsibility is more fundamental than *mens rea* in the sense of knowledge of circumstances or foresight of consequences; for even where *mens rea* in that sense is not required and responsibility is 'strict' or 'absolute' (as it is said to be, e.g. in the case of dangerous driving), this element, according to some modern writers, is still required.

I am doubtful about this doctrine on two quite distinct scores. First, I cannot find in any legal writings any clear or credible account of what it is for conduct to be voluntary and not involuntary in the sense required: secondly, I am very doubtful whether the doctrine that this element is required, even in cases of strict or absolute liability, is one which the courts at present accept.

The performance of a human action is a very complex affair involving the co-presence and the co-ordination of many different elements.[1] It may go wrong in many different

[1] See, for an illuminating account of the complexity of human action, J. L. Austin, 'A Plea for Excuses' *Proceedings of the Aristotelian Society,* Vol. 57 (1956-7), p. 1.

ways, and some of these are brought home to us in such melancholy adages as 'accidents will happen' or 'we all make mistakes'. If we are to succeed in doing such simple things as strike a match, kick a football or write down a simple sentence, we must have a capacity to control the movement of our limbs in certain ways, we must be able to recognize certain objects for what they are, and we must have a certain knowledge or foresight of the consequences of manipulating them or interfering with them. If we consider the types of action with which the criminal law is most concerned, e.g. killing or wounding others, it is plain that these may be done on the one hand by someone possessed of full knowledge that he is doing these things, or, on the other hand, by one who fails to foresee the relevant consequences of his movements or who lacks some relevant knowledge concerning the circumstances in which he is placed or concerning the character of the things or persons affected by his movements. Smith has indeed shot Jones; but fuller inquiry shows that he did it unintentionally: perhaps he shot at a bird not foreseeing that Jones would suddenly and without warning step into the line of fire, or that the bullet would ricochet off a tree: or perhaps he thought on good grounds that the gun he was playfully pointing at Jones was unloaded. These are the 'accidents' and 'mistakes' of our two melancholy adages, and it is the knowledge and foresight absent in such cases to which most Anglo-American lawyers refer when they speak of *mens rea*.[2]

Undoubtedly there is another kind of defect in human conduct which is different from such lack of knowledge or foresight, and which may well seem a far more fundamental defect than these. Where this defect is present, the movements

[2] But the terminology in the books is not settled. J. W. C. Turner uses *mens rea* to include *both* the element which makes conduct 'voluntary', and foresight of consequences: see Kenny, *Outlines of Criminal Law* (19th edn.), pp. 29, 30, and 'The Mental Element in Crimes at Common Law' in *The Modern Approach to Criminal Law*, pp. 203, 205. See also Cross and Jones, *Introduction to the Criminal Law* (5th edn.), pp. 30–35. Glanville Williams, while distinguishing 'the requirement of will' from *mens rea*, criticizes the decision in *R.* v. *Harrison-Owen*, (1951) 2 All E.R. 726 in which this distinction was considered relevant to the admissibility of similar facts. (*Criminal Law, The General Part* (2nd edn.), pp. 11–15.)

of the human body seem more like the movements of an inanimate thing than the actions of a person. Someone, unconscious in a fit of epilepsy, hits out in a spasm and hurts another; or someone, suddenly stung by a bee, in his agony drops and breaks the plate he is holding. A layman might say that in these cases the man's movements were 'involuntary' or 'not under his control' and if we call these 'actions' it is only in the thinnest of all senses of that wide word, i.e. the sense in which it embraces anything we can say by putting together a verb with a personal subject. Except in this virtually all-embracing sense, the layman, like the lawyer, would wish to distinguish tumbling downstairs from walking downstairs as not 'really' an action at all. Many English legal writers, under the influence of the inherited theory which I shall expound later, say that where conduct is defective in this very fundamental way there is no 'act', although there are movements of the body; for an 'act' is something more than such a movement. This way of putting it at least serves to mark off the fundamental character of such defects from others where the ordinary requirements of *mens rea* are not satisfied, and it strikingly suggests the idea that what is missing in such cases is a minimum link between mind and body, indispensable for any form of criminal responsibility.

Hill v. *Baxter*

A much discussed recent case has been said to illustrate the doctrine that this minimum link between mind and body is required even in cases where responsibility is 'strict' or 'absolute'. The offence vulgarly called 'shooting the lights' is one form of an offence defined more sedately by section 49(b) of the Road Traffic Act 1930 as failing 'to conform to the indication given by the [traffic] sign.' In *Hill* v. *Baxter*[3], which

[3] (1958) 1 Q.B. 277; (1958) 1 All E.R. 193. This case is elaborately discussed in Prevezer, 'Automatism and Involuntary Conduct', *Crim. L.R.* (1958), pp. 361, 440 and Edwards, 'Automatism and Criminal Responsibility', *21 M.L.R.*(1958), p. 375.

reached the Divisional Court in December 1957, on a case stated by justices, a driver of a van was charged with an offence under section 49 (b) and with dangerous driving under section 11(1) of the Act. He had, according to the evidence, driven his car at a high speed across the road junction where there was an illuminated 'Halt' sign and collided with a car and then overturned. The accused pleaded that he was not responsible under these sections, because he had become unconscious and remembered nothing from some time before reaching the crossing until after the collision. This plea was accepted by the justices on the footing that the loss of memory could only be attributed to the accused being overcome by illness without warning. The Divisional Court, however, held that the accused had not tendered sufficient evidence that he was in a state of automatism or abnormal unconsciousness as distinct from ordinary sleep. He might just have felt drowsy and fallen asleep. So they sent back the case with a direction to convict, on the ground that the accused had failed to discharge the onus of proof. But though the plea failed, the judges made two important observations about the law. First Lord Goddard, then Lord Chief Justice, said 'the first thing to be remembered is that the Road Traffic Act, 1930, contains an absolute prohibition against driving dangerously or ignoring "Halt" signs. No question of *mens rea* enters into the offence; it is no answer to a charge under these sections to say: "I did not mean to drive dangerously" or "I did not notice the 'Halt' sign"'.[4] Secondly both Lord Goddard and Pearson J. agreed that there may be some states of unconsciousness, such as those due to a stroke or epileptic fit which, if satisfactorily proved, would exclude liability for dangerous driving even though it is an offence of 'absolute prohibition.' Pearson J. thought that liability might be excluded not only in cases of stroke or epilepsy but also if the driver was stunned by a blow on the head from a stone, or if he was attacked by a swarm of bees 'so that he is for the time being disabled and prevented from exercising any directional control over the vehicle, and any

[4] (1958) 1 Q.B. 277, at p. 282.

movements of his arms and legs are solely caused by the action of the bees.'[5] He drew attention, however, to the possibility that if a man drove knowing that he was liable to have an epileptic fit his driving in such circumstances might be considered dangerous. In a somewhat similar case, *R.* v. *Sibbles* tried at Leeds Assizes,[6] where a driver was charged with causing death by dangerous driving, it was argued in his defence that he had driven in a state of automatism. The Judge, Paull J., there directed the jury that they should convict 'unless they found that the defendant suddenly and unexpectedly was deprived of all thought and that deprivation was not connected with any deliberate act or deliberate conduct of his, and arose from a cause which a reasonable man would have no reason to think and the defendant did not think might occur.'[7]

It seems plain from these cases that even where liability is strict there are certain forms of unconsciousness and certain types of failure of muscular control which will exclude liability, though perhaps only if the accused could not have foreseen their occurrence. But it is by no means clear to me that this shows that the courts adhere to any general doctrine that a voluntary 'act' is required for all criminal liability, even where liability is 'strict' or 'absolute'. None of the judges in *Hill* v. *Baxter* based their observations concerning the possible relevance of epilepsy, strokes, bee stings or stunning blows from stones on any such doctrine; but instead they appealed to quite different considerations which I shall discuss later and which may indeed prove to be inadequate in other cases. Yet in his important article on Automatism and Criminal Responsibility, Mr. J. Ll. J. Edwards speaks of the 'voluntary act requirement' as 'the fundamental requirement of all criminal liability, whether the offence is one of absolute prohibition or one involving proof of a guilty mind and whether statutory or common law in origin. This requirement, stated in its

[5] Ibid, at p. 286.
[6] On 13 July 1959. See *Crim. L.R.* (1959) p. 660.
[7] Ibid, at p. 661.

simplest form, is that the "act" of the accused, in the sense of a muscular movement, must be willed. It must be a voluntary expression of the accused's will. . . .'[8]

The Meaning of the General Doctrine

For such a view Mr. Edwards has certainly ample support in what is said by English writers on the criminal law. In many textbooks there are general assertions that for *all* criminal responsibility conduct must be 'voluntary',[9] 'conduct [must be] the result of the exercise of his will'[10] there must be an 'act with its element of will,'[11] 'an act due to the deliberate exercise of the will.'[12] Yet, surely, even if there is any such general doctrine, these phrases are very dark. What does the doctrine mean? What after all *is* the will? If we search the books on criminal law for an answer to this question we will find two things to help us. The first of these is a list of examples of cases in which this element is lacking. Nearly all these examples are hypothetical ones, either suggested by the writer, or like those in *Hill* v. *Baxter* referred to as possible cases by judges, but distinguished from the actual cases before them. It is, I think, profitable in considering these textbook examples to divide them into two classes, those where the subject is conscious and those where he is unconscious.

(i) Conscious.

(*a*) *Physical compulsion of one person by another*. A holds a

[8] 21 M.L.R., pp. 379–80 and footnotes. He notes that *R.* v. *Larsonneur* (1933), 24 Cr.App.R. 74 has been cited as the solitary exception to 'this principle that all criminal liability is based on proof that the accused's "act" was voluntary', (p. 379, n. 22).

[9] Turner in Kenny (19th edn.), pp. 29, 46. He excepts cases such as *R.* v. *Larsonneur* (*supra*) where a statute is so worded as necessarily to exclude such a requirement. (pp. 30 n. 3; 46 n. 2).

[10] Ibid, p. 27.

[11] Glanville Williams op. cit., pp. 13, 14. He observes that the importance of the doctrine is said to relate to crimes of strict responsibility but he finds it 'hard to imagine a practical case' (ibid, p. 14). He also notes the difficulty of applying the doctrine to inadvertent omissions (ibid, p. 15).

[12] Harris, *Criminal Law* (19th edn.), pp. 20, 21. For the proposals of the American Law Institute's *Model Penal Code* Art, 2. s. 2.01 and comment requiring a 'voluntary act', see Edwards, loc. cit. p. 379, n. 22.

weapon and B against A's will seizes his hand and therewith stabs C.[13]

(*b*) *Muscular control impaired by disease.* A is afflicted by St. Vitus dance. Harm results from his uncontrolled movements.[14]

(*c*) *Reflex muscular contraction.*[15] A while driving a car is attacked by a swarm of bees or hit by a stone and the car is put temporarily out of his control.

(ii) Unconscious.

(*a*) *Natural sleep at normal time.*[16] A woman while sleep-walking takes an axe and kills her daughter.

(*b*) *Drunken stupor.* A woman in a drunken stupor overlays and kills her child.[17]

(*c*) *Sleep brought on by fatigue.* A driving a car home from night shift falls asleep and runs into a detachment of soldiers.[18]

(*d*) *Loss of consciousness involving collapse.*[19] A is suddenly deprived of consciousness and collapses owing to a stroke or epilepsy.

(*e*) *Automatism or abnormal state of unconsciousness not involving collapse.* A enters a dwelling house at night in a state of somnambulism or 'automatism'.[20]

Here, then, is a list of very interesting, though mainly hypothetical, cases of abnormalities in human conduct. We can all see that something is far more fundamentally wrong than in those cases where a subject who is conscious simply does something by mistake, or without foreseeing the

[13] Kenny, op. cit., pp. 29 (case cited from Hale P.C. 434).

[14] Kenny, op. cit., p. 29.

[15] Hypothetical cases from dicta in *Kay* v. *Butterworth* (1945), 173 L.T. 191; 61 T.L.R. 452 and *Hill* v. *Baxter* (*supra*), p. 92.

[16] Hypothetical case in Glanville Williams, op. cit. p. 13, based on Australian case cited (ibid, n. 6). Kenny op cit., pp. 29, 30.

[17] Glanville Williams, op. cit., p. 13. Kenny, op. cit., pp. 29, 30.

[18] *Kay* v. *Butterworth:* where the accused was convicted of dangerous driving and of driving without due care and attention contrary to s.12 of the Road Traffic Act 1930, on the footing that he must have earlier known that drowsiness was overtaking him and yet continued to drive. This approach is similar to that of Pearson J. in *Hill* v. *Baxter* (discussed *infra*, pp. 108–11) and suggests that driving while asleep can never be a 'voluntary act'.

[19] Dicta in *Kay* v. *Butterworth* and *Hill* v. *Baxter.* Edwards, op. cit. p. 300.

[20] *R.* v. *Harrison-Owen* (*supra*), p. 91, n.2.

consequences. But what precisely is it that is wrong in these cases? What common feature have they leading to their classification as cases where there is no 'act of will' or no 'expression of the will' or no 'voluntary conduct'? Or to put the same point in a different way, what is it that is present in normal action which makes it a satisfactory example of 'voluntary conduct' or 'willing', etc? Such intelligible answers as the law books give to this question consist in the remnants of a theory as to the nature of human action which goes back at least to the eighteenth century. It is expounded clearly by Austin in Lectures XVIII-XIX[21] of his Lectures on Jurisprudence, but was derived by him from Dr. Thomas Brown[22], poet, philosopher and physiologist in Edinburgh in the first years of the last century. The theory is simply this: a human action is strictly speaking merely a muscular contraction. The usual terminology of ordinary speech—the verbs of action like 'shooting', 'killing', 'hitting'—are inaccurate and misleading, because they misrepresent as single actions what in fact are combinations of muscular movements and later consequences. We should, therefore, confine the word 'act', if we are to think and speak scientifically and clearly, to the mere muscular contraction. This is the first element of the theory. The second is that an 'act' is not *just* a muscular contraction but one which has a special psychological cause. It is caused by a pre-existing desire, which Austin called a 'volition' or 'act of will', for the muscular contraction. Here is the dividing line between mere involuntary movements, like tumbling downstairs, and voluntary actions, like walking downstairs. In the one case the muscular contractions are desired, and caused by the desire for them, and in the other they are not. This is the minimum, indispensable connexion between mind and body if there is to

[21] 5th edn. 1885: pp. 411–24.

[22] Austin cites Brown's *Enquiry into the Relation of Cause and Effect* (1818) Part 1, s.3 and says 'he was (I believe) the first who understood what we would be at, when we talk about the *will*, and the *power or faculty of willing*' (ibid, p. 412). *Brown* (1778–1820) was one of the first contributors to the *Edinburgh Review*, studied both law and medicine, practised the latter and shared the Chair of Moral Philosophy at Edinburgh University with Dugald Stewart. He was much influenced by Hume.

be an 'act' and responsibility. Of course in a full blown action (according to ordinary speech) like 'killing' there is, if it is done intentionally, besides the desire or 'volition' for the muscular movement, knowledge of circumstances and foresight or desire of consequences, and in criminal cases these elements may also be necessary for responsibility as part of *mens rea*. But these are to be distinguished from the 'volition' or 'act of will' which is solely a desire for the muscular movements.

Let me quote some passages from Austin's own account of this theory which has inspired all the subsequent, but far less clear, discussions of 'acts' in books on criminal law. He says:

> 'Certain movements of our bodies follow invariably and *immediately* our wishes or desires for those *same* movements: Provided, that is, that the bodily organ be sane, and the desired movement be not prevented by any outward obstacle. . . . These antecedent wishes and these consequent movements, are human *volitions* and *acts* strictly and properly so called. . . . And as these are the only *volitions*; so are the bodily movements, by which they are immediately followed, the only *acts* or *actions* (properly so called). It will be admitted on the mere statement, that the only objects which can be called acts, are consequences of Volitions. A voluntary movement of my body, or a movement which follows a volition, is an *act*. The *in*voluntary movements which are the consequences of certain diseases, are *not* acts. But as the bodily movements which immediately follow volitions, are the only *ends* of volition, it follows that those bodily movements are the only objects to which the term 'acts' can be applied with perfect precision and propriety. . . . Most of the names which seem to be names of acts, are names of acts, *coupled with certain of their consequences*. For example, if I kill you with a gun or pistol I *shoot* you: And the long train of incidents which are denoted by that brief expression, are considered (or spoken of) as if they constituted an *act* perpetrated by me. In truth, the only parts of the train which are my act or acts are the muscular motions by which I raise the weapon; point it at your head or body, and pull the trigger. These I *will*. The contact of the flint and steel; the ignition of the powder, the flight of the ball towards your body, the wound and subsequent death, with the numberless incidents included in these, are *consequences* of the act which I *will*. I *will* not those consequences, though I may *intend* them.'[23]

Here then is an agreeably simple answer to our question. We know now what the general doctrine means: it defines an act in terms of the simplest thing we can do: this is the

[23] op. cit. passages from Lecture XVIII, pp. 411, 412, 414, 415.

minimum feat of contracting our muscles. Conduct is 'voluntary' or 'the expression of an act of will' if the muscular contraction which, on the physical side, is the initiating element in what are loosely thought of as simple actions, is caused by a desire for those same contractions. This is all the mysterious element of the 'will' amounts to: it is this which is the minimum indispensable link between mind and body required for responsibility even where responsibility is strict.

It would be interesting to trace in detail the descent[24] of this doctrine to the modern writers on the criminal law whom I have quoted. I cannot do that here; I will however observe that the terminology in which the doctrine is expressed has, in the course of the descent, become very much less precise. For later authors do not ever plainly say that the psychological element which makes conduct 'voluntary' is just a desire for muscular contractions. Instead we are told that it is 'an element of will' or 'operation of the will' (from which muscular contractions 'result') or 'a mental attitude to conduct' as distinct from a 'mental attitude to the consequences of conduct'.[25] But in spite of this laxer terminology, it seems clear that substantially the same doctrine is intended.[26]

Austin's doctrine of the act and the volitions which make it voluntary is then very simple, and has stood the test of time in the sense that it is still to be found in our law books. But is it right? There are, I think, at least two reasons why it cannot intelligibly or correctly characterize, as it is supposed to do, a minimum indispensable connexion between mind and body present in all normal action, and generally required for responsibility. The first reason is that though the doctrine is

[24] Some of the principal stages are Holland: *Elements of Jurisprudence* (6th edn.) p. 93: Markby: *Elements of Law* (6th edn.), p. 116: *Clark*: *Analysis of Criminal Liability*, p. 23; J. F. Stephen: *A General View of the Criminal Law of England*, chap. v; Holmes: *The Common Law* pp. 54, 91; W. W. Cook, *26 Yale Law J.* (1917), p. 645. For a similar doctrine among philosophers, see H. A. Prichard, *Moral Obligation: Essays and Lectures*, p. 19. The same doctrine is criticized in Wittgenstein: *Philosophical Investigations*, pp. 159–62, and G. E. M. Anscombe: *Intention*, pp. 53, 54.

[25] Kenny, op. cit., *supra* p. 95, n. 9 at p. 30.

[26] Glanville Williams, op. cit., p. 12 refers to 'Holmes (following Austin).'

said by modern writers to apply to omissions[27] (e.g. failing to conform to a traffic sign) as much as to positive interventions, it is surely absurd even to attempt to fit omissions into such a picture of voluntary or involuntary conduct. For the doctrine defines what is involuntary as a muscular *contraction or movement* which occurs without the preceding volition or desire for it. At the best, this only makes sense if applied to uncontrolled involuntary interventions; these are indeed involved in, e.g. knocking over a vase during an epileptic fit. But where someone owing to sudden descent of a paralysis or a coma simply *fails* to do something which he is required to do (e.g. stop at a traffic signal) we cannot express what is defective by saying that a muscular movement or contraction has occurred without a desire for it; *ex hypothesi* in the case of omissions no muscular movement or contraction need occur. The theory therefore only tells us when a positive intervention is involuntary and gives us no criterion for saying when an omission is involuntary. Moreover we cannot rescue the theory from this difficulty by amending it generously to mean that omissions are voluntary if the *failure* to contract the muscles[28] so as to do the action required was caused by a desire *not* to contract the muscles and involuntary if it was not so caused. This would have very unwelcome consequences for legal responsibility: for the only omissions which would then be culpable would be deliberate omissions. We could then only punish those who failed to stop at traffic lights if they deliberately shot the lights. Yet as is clear from *Hill* v. *Baxter*, it is certainly the law that at least some forms of non-deliberate *inadvertent* omission to conform to a traffic signal are punishable; and generally throughout the law, we would surely wish to distinguish the inadvertent omission of the ordinary healthy man from the omission of the man suddenly paralysed or suffering from a stroke. Yet the theory cannot help us to make this distinction

[27] Kenny, op. cit., pp. 17, n.1 and 30, n.3. '*mens rea* in this sense relates not to the harm which the man brought about but to the movements (or *omissions*) by which he brought it about' (ibid, p. 30: my italics).

[28] Glanville Williams observes (op. cit., p. 16) that in a 'mere negligent omission . . . a man . . . has not acted' and that here 'it is difficult to find an "act" in any meaning of the term'.

for, in neither case, is there any 'volition' or desire to make (or, in the amended version, to omit) muscular movements.

So much for the first objection. But even where the theory looks as if it might work, even where it seems most at home, namely in the cases of involuntary *movements* such as are made in epileptic seizures or in reflex responses to bee stings, its account of human action is really nothing more than an outdated fiction—a piece of eighteenth-century psychology which has no real application to human conduct. This is one reason why the talk of 'volitions', 'acts of will' causing 'muscular contractions' is so rarely found in either the language of ordinary people or the courts. For the theory splits an ordinary action into three constituents: a desire for muscular contractions followed by the contractions, followed by foreseen consequences. Such a division is quite at variance with the ordinary man's experience and the way in which his own actions appear to him. This surely is a fatal defect in any account of action supposed to help us to characterize the mental conditions required for the ordinary man's responsibility.

In making this objection I am not of course denying that there is a defect in human conduct present in the cases we have listed which is a different defect from lack of knowledge or foresight and a more fundamental one. My claim is only that we cannot convey the difference between the normal case and these very abnormal ones, by saying that in the normal case there is a desire for the muscular contractions which is absent in the abnormal case. For the desire for muscular contraction as a component of ordinary action is a fiction. To show this I shall make a short excursion into the realm where philosophy and psychology meet, and my excuse for this is that the very doctrine we are considering is itself a misleading antiquated piece of philosophical psychology.

The relevant points are not very recondite. The first of them is that a desire to contract our muscles is a very rare occurrence: there are no doubt *some* special occasions when it would be quite right to say that what we are doing is contracting our muscles, and that we have a desire to do this. An example of this is what we may do under instruction in a gymnasium. The instructor says 'lift your right hand and contract the muscles of the upper arm'. If we succeed in doing this (and it is not so easy) it would be quite appropriate to say we desired to and did contract our muscles. In *this* case 'I contracted my muscles' would be a sensible answer to the question 'What did you do?' I draw attention to this not as a matter of language, but because language here does usefully mark a vital, factual distinction which the theory we are criticizing ignores. Another example is the situation when we are baffled in the physical effort to execute some action. The door handle will simply not turn when we try to turn it in the ordinary way; so we clench it with a special grip, and here we may actually be conscious of the muscles we must contract and actually desire to contract them. But these *are* special occasions, and in such cases the whole outward posture of our body, and our concentration of gaze on the parts of our body which we are intent on moving would usually show this to an outsider. Similarly the inward experience of the actor in such actions with its special concentration of attention on muscular movement would also be different from ordinary occasions when we do actions. When we shut a door, or when we hit someone, or when we fire a gun at a bird, these things are done without any previous thought of the muscular movements involved and without any desire to contract the muscles. No doubt sometimes we may previously deliberate about doing these actions, and we may then have some image of ourselves doing these things or of the final result: we may see 'in the mind's eye' the opened door or our victim's bleeding nose and we may desire this; but this is not a desire or awareness of our muscular movements. The simple but important truth is that when we deliberate and think about actions, we do so not in terms of muscular movements but in the ordinary terminology of actions. Of course muscular movements are *involved* in all such actions; but that does not show either that we are aware of them before acting or that we have a desire for them.

The same point, viz. that this eighteenth-century, atomistic account of action misrepresents the way in which actions appear to ordinary men in doing ordinary actions, may be made in another way. If we are given a simple order, e.g. to write down the letter 'Q' or to kick a football or to say the word 'Equity' we can, if normal, comply quite easily. But if someone says 'Don't actually *do* these things but tell me what muscles you have to contract to do them' this is quite another (and a very difficult) thing for anyone who is not a trained physiologist to do. It is of course not impossible for us to *find out* what muscles are contracted in doing these things, but it is important to notice how we should get this information. In order to find out the facts we should first imagine ourselves complying with the order, i.e. writing down the letter 'Q', kicking the football, or saying 'Equity' under our breath and then we could try to see or feel what muscles were moved in the process. This is quite a sophisticated experiment and surely shows that our *primary* awareness of our own actions is not that of a physiologist: it does not include a knowledge of the muscular movements required, and *a fortiori* does not include a desire for them. What happens in normal action is that if we decide to do something we think of it in the ordinary terminology of action (as hitting someone, or writing something) and given that we have learnt to do these things and our faculties are unimpaired, our muscular movements normally follow smoothly on our decision. We do not have to launch our muscles into action by desiring that they contract as the Austinian terminology of 'acts' caused by 'volitions' suggests. Of course, we can tone down this theory to make it more plausible: if the theory that in all normal action there is present a desire to contract the muscles is taken to mean merely that when we act we desire to do some action (e.g. hit someone) which *involves* muscular contractions, this is no doubt broadly acceptable, if not very informative. But to tone the theory down in this way is not only to depart from at any rate Austin's clear meaning; it also involves abandoning the idea, still treasured by some, that if we confine the word 'act' to muscular contractions we are speaking more accurately or scientifically than if we use the ordinary verbs of action. For on the toned-down version of the theory, the desire to contract our muscles is not something the occurrence of which we independently observe or can verify: it is merely 'conclusively inferred' from the fact that we desire to do some actions in the ordinary 'loose' sense of action: so it presupposes the ordinary man's ordinary description of what he does and desires to do in terms, not of muscular contractions, but of such things as kicking a ball, hitting a man, or writing a letter.

So, neglecting the toned-down version of the theory, I shall summarize this minor excursion into the philosophy and psychology of action by saying that the eighteenth-century theory that has got into our law books through Austin is first, nonsensical when applied to omissions, and secondly cannot characterize what is amiss even in involuntary interventions; for the desire to move our muscles, which it says is missing there, is not present in normal voluntary action either.

The General Doctrine Reconstructed

Most people, lawyers and laymen alike, would I think agree that in our list of examples of involuntary conduct (conscious and unconscious), some radical defect is present, and some vital component of normal action is absent, even if Austin's terminology of 'desire' or muscular movement or volitions misdescribes it. For the cases do not seem to be a *mere* list, without any unifying feature to justify treating them alike as cases where conduct is not voluntary. If it is the policy of the law to mark these cases off, there seems some good factual basis for this policy. Is it then possible to give a more adequate account than that of the traditional theory? Or must we leave the dark phrases 'not governed by the will', 'no act of will', 'involuntary', 'no operation of the will', &c. unexplained?

In fact, I think it would not be difficult to construct an account which would explain and justify the intuitive feeling that, in all these cases, there is some more fundamental defect than lack of knowledge or foresight. By a 'more adequate' account I mean one which involves no fictions; which is better fitted to the facts of ordinary experience; and which could be used by the courts in order to identify a range of cases where the requirement of a minimum mental element for responsibility is not satisfied. Such an account could cover both the conscious and unconscious examples suggested in the books, but it would necessarily differ from the kind of general explanation given there in two main ways. First it would be disassociated from any claim that the ordinary way of talking about actions was inferior to, or less accurate than, the definition of acts as muscular contractions. Secondly, omissions would have to be treated separately from positive interventions. Granted these two things, we could then characterize involuntary movements such as those made in epilepsy, or in a stroke, or mere reflex actions to blows or stings, as movements of the body which occurred although they were not appropriate, i.e. required for any action (in the ordinary sense of action) which the agent believed himself to be doing. This, I think, reproduces what is in fact meant by ordinary people when they say a man's bodily movements are uncontrolled, as in the case of a reflex or St. Vitus dance. Such movements are 'wild' or not 'governed by the will' in the sense that they are not subordinated to the agent's conscious plans of action: they do not occur as part of anything the agent takes himself to be doing. This is the feature which the Austinian theory represents in a distorted form by identifying the involuntary movements as those which are not caused by a desire for them.

In the unconscious cases, e.g. of epilepsy, automatism, etc., the same test can be used. Here too, the movements which we call involuntary are not part of any action the agent takes himself to be doing, because, being unconscious, he does not take himself to be doing any action. This test, it should be noted, preserves the distinction between involuntary conduct and mere lack of knowledge of circumstances or foresight of consequences, and so reproduces the sense that we have, in involuntary movements, a different and more fundamental defect. For one who merely fails to foresee that the gun he fires will harm someone still makes voluntary muscular movements, i.e. movements appropriate to the action of firing the gun, which he knows he is doing; whereas the involuntary tremors of the palsied man, who breaks a glass, are appropriate to no action which he believes himself to be doing.

Omissions must, I think, be catered for separately, though this can and should be done in a way which reveals that their voluntary or involuntary character depends on the same general principle as positive interventions. When a man fails to do some positive action demanded by the law, his failure to act is involuntary if he is unconscious and so *unable* to do any conscious action, or if, though conscious he is *unable* to make the particular muscular movements required for the performance of actions demanded by the law. In the case of omissions it is this inability which the Austinian theory misrepresents as an absence of desire for muscular movements. Plainly *abilities* and *desires* are different, and the latter seem irrelevant here.

These two related criteria, one for involuntary movements and the other for involuntary omissions, characterize, without bringing in the fictitious desire for muscular movements, different aspects of a single fundamental defect: viz. a man's lack of conscious control over his muscular movements. To summarize the point in crude terms, we may say that the controlling agency is not a desire for muscular movements but the mind of a man bent on some conscious action: control may be lacking for different reasons: (1) because the controlling agency is 'out of action' (as in movements or omissions where the agent is unconscious), (2) because either the muscles to be controlled, though healthy, move in ways not required for any conscious action (involuntary movements) or through some disease or defect are incapable of moving as and when required for conscious action (involuntary omissions where the agent is conscious).

The General Doctrine and the Courts

That the general doctrine could be more intelligibly restated, in some such way as I have indicated, is clear: and I think it is also clear that, thus restated, it would fairly well reproduce or make explicit the common conception that in these cases of unconsciousness, automatism, reflexes and the like there is a far more serious abnormality than mere lack of knowledge of circumstances or foresight of consequences. What is missing in these cases appears to most people as a vital link between mind and body; and both the ordinary man and the lawyer might well insist on this by saying that in these cases there is not 'really' a human action at all and certainly nothing for which anyone should be made criminally responsible however 'strict' legal responsibility may be.

That the courts ought to accept this doctrine in order to accord with common convictions or in order to do justice may be clear. Yet when we turn from the books to the cases, it seems to me far from clear that they do accept it. Of course, cases where this point has to be considered are always likely to be few and far between. The reasons for this are, first, the simple fact that the abnormal states of mind and body (St. Vitus dance, automatism, epilepsy, uncontrolled reflexes) which, in accordance with the general doctrine, make action involuntary, are mercifully rare. Certainly they are much rarer than the ordinary mistakes or accidents which exclude intention or *mens rea* in the ordinary sense. Secondly, even where we have a case of abnormal involuntary movement it may be quite unnecessary to consider or apply the general doctrine that this, as such, excludes responsibility; for in most cases of importance *mens rea* in the sense of knowledge of circumstances and foresight of consequences is an essential element in responsibility. The unconscious epileptic or the man who in his sleep kills another, will certainly lack such knowledge and foresight; so too will the man who errs because he is a victim of St. Vitus dance, or is suddenly stung by bees. Hence in most cases the lack of knowledge or foresight will itself be enough to exclude liability, and it will not be necessary to bring in any doctrine concerning involuntary movement even in a case where this far more fundamental defect is also present.[29]

It follows that this doctrine is only important in the criminal law where responsibility, as in the motoring cases, is 'strict'[30]

[29] None the less, even where, as in a murder, *mens rea* is required the courts have sometimes discussed the question whether the accused's 'conduct' was voluntary or not: see *Fain* v. *Commonwealth* (1879), 78 Ky 183, 39 Am. Rep. 213.

[30] Or possibly where negligence is the basis of liability. Some writers exclude negligence from the scope of the term '*mens rea*'.

i.e. where *mens rea* in the sense of knowledge of circumstances or foresight of consequences is not required. Yet if you look at the few cases which we have, it is clear that the judges do not talk on this topic the language of the books. Not only do they not refer to muscular contractions or 'volitions' or desires for them but they do not speak as if they were faced with any general doctrine that, however strict liability may be, voluntary movements or omissions are still necessary for responsibility. Instead they discuss the meaning of the words in the statutes which they are considering, e.g. words like 'driving' used in section 11 of the Road Traffic Act 1930 making driving dangerously an offence. Thus in *Hill* v. *Baxter* Lord Goddard's view that responsibility might be excluded in some of our cases was expressed thus: 'I agree that there may be cases where circumstances are such that the accused *could not really be said to be driving at all.* Suppose he had a stroke or an epileptic fit, both instances of what may properly be called acts of god; he might well be in the driver's seat even with his hands on the wheel, but in such a state of unconsciousness that *he could not be said to be driving*. . . . A blow from a stone or swarm of bees introduces I think some conception akin to *novus actus interveniens*.'[31]

Pearson J. who made it even plainer that he conceived of himself as concerned with construing the word 'drive' in section 11 of the Act, said 'in any ordinary case, when once it has been proved that the accused was in the driving seat of a moving car, there is, prima facie, an obvious and irresistible inference that he was driving it. No dispute or doubt will arise on that point unless and until there is evidence tending to show that by some extraordinary mischance he was rendered unconscious or otherwise incapacitated from controlling the car. Take the following cases:

(i) The man in the driving seat is having an epileptic fit, so that he is unconscious and there are merely spasmodic movements of his arms and legs.

(ii) By the onset of some disease he has been reduced to a state of coma and is completely unconscious.

[31] (1958) 1 Q.B. 277, at p. 283 (my italics).

(iii) He is stunned by a blow on the head from a stone which passing traffic has thrown up from the roadway.

(iv) He is attacked by a swarm of bees so that he is for the time being disabled and prevented from exercising any directional control over the vehicle and any movements of his arms and legs are solely caused by the action of the bees.

In each of these cases it can be said that at the material time he is not driving and, therefore, not driving dangerously.

Then suppose that the man in the driving seat falls asleep. After he has fallen asleep he is no longer driving, but there was an earlier time at which he was falling asleep and therefore failing to perform the driver's elementary and essential duty of keeping himself awake and therefore he was driving dangerously.'[32]

Of course, both the general doctrine, and this way of approaching the matter *via* the meaning of the words used in defining the offence, would come to the same thing *if* it were the case that, whenever we have an active verb like 'drives', this implies, as part of its meaning, the existence of the minimum form of conscious muscular control, upon which the general doctrine insists. As a matter of ordinary English this is however not the case. The phrase 'sleep-walking' is alone sufficient to remind us that if the outward movements appear to be co-ordinated as they are in normal action, the fact that the subject is unconscious from whatever cause does not prevent us using an active verb to describe the case, though we would qualify it with the adverb 'unconsciously', or with the adverbial phrases 'in his sleep', 'in a state of automatism', etc. So in the case of 'driving' it would be natural, as a matter of English, to distinguish those cases where the movements of the body are wild or spasmodic or where the 'driver' simply slumps in his seat or collapses over the wheel, from cases where, though unconscious, he is apparently controlling the vehicle, changing gears, steering, braking, etc. In the latter case it might well be said that he drove the vehicle, changed

[32] (1958) 1 Q.B. 277, at p. 286.

gear, braked, etc. 'in his sleep' or 'in a state of automatism.' Such cases can certainly occur.

It is true that in *Hill* v. *Baxter*, Pearson J. did not make this distinction but said quite generally 'after he has fallen asleep he is no longer driving'. His view was apparently that, in all such cases, the dangerous driving consisted not in anything which happened while the driver was unconscious, but in the earlier driving in a drowsy condition or 'in failing to perform the driver's duty of keeping himself awake.' Perhaps Pearson J.'s apparent refusal to distinguish the case where we would say 'he drove the car in his sleep' from the case where we would say 'he was fast asleep in the driving seat, not driving at all' is some slight indication that judges, in spite of claiming to do so, do not really view the question simply as one of the meaning of the words of the statute and quite independent of the general doctrine. Perhaps they are influenced in construing the words by the general doctrine, though this is left unexpressed.

In any case it seems clear that if the question of responsibility were really to be settled simply by reference to the question whether or not the accused's conduct could, in accordance with English usage, be described as 'driving', this might have very unfortunate results. For it might well happen that a driver became unconscious though he had never done so before, and had no warning of or reason to suspect an onset of this condition. His unconscious conduct might take that outwardly co-ordinated form which we might well describe as 'driving while unconscious'. If the applicability of this phrase settled the matter he would be responsible under section 11 of the Act, whereas a person who collapsed in the seat unconscious would not be responsible. Yet surely no moral consideration or social policy could justify a distinction here. Perhaps some such consideration rather than any general doctrine is likely to lead the courts to adopt the view implied in Pearson J.'s approach, viz. that from the moment a driver becomes unconscious (whether or not he collapses) he is no longer 'driving' (in law) but he will be responsible if at some earlier stage he drove consciously with the knowledge of his propensity to lose consciousness.

This approach, suggested in *Hill* v. *Baxter* and in substance adopted in *R.* v. *Sibbles* involves, as it were, ante-dating the commission of the offence of dangerous driving to the time when the accused was conscious. It does involve some straining of the English language; for a man who is driving impeccably so far as concerns the conduct of the vehicle in the conditions of traffic, etc. on the road is oddly said to be driving in a *manner* dangerous to the public even if he knows he is going to fall asleep or have a fit. He would no doubt be driving in a *condition* dangerous to the public; and this indeed would be true even if he did not know of his condition. But it is too late to protest against what is, after all, a very mild and quite salutary stretching of words. More important is the fact that this antedating approach is not possible in other cases where liability is strict and where the courts would certainly wish to convict a person who was asleep at the time he did what the law forbids, if he knew of his propensity to fall asleep. Take for example, the other charge in *Hill* v. *Baxter* under section 49(b) of the Road Traffic Act, failing 'to conform to the indication given by the [traffic] sign.' This offence cannot be handled in the same way, though in *Hill* v. *Baxter* the court thought the accused guilty of it even if he was asleep at the time he reached the lights. This cannot very well be reconciled with any general doctrine requiring that the accused, if he is to be convicted, must have been conscious at the time he committed the offence; for even if it was the case that, before reaching the lights, the driver knew or believed he was going to fall asleep, or otherwise lose consciousness, we could not, however much we stretch English, say that it was *then* that he failed to conform to the traffic sign, as we could say he *then* drove in a dangerous manner (or condition). You cannot cross your bridges before you come to them; equally, you cannot shoot your lights before you come to them.

On what theory is it then that we hold the unconscious driver responsible for such offences? And how can we, in such cases, distinguish, as we would no doubt wish to do, between the unconscious man who shoots the lights having driven with knowledge that he was going to lose consciousness and the unconscious man who does the same, but had no reason to suspect a loss of consciousness? If we insist that it is simply a matter of construing the words 'fails to conform' to a traffic sign, probably both must be guilty; for it is not very plausible to argue that 'fails to conform' as a matter of English demands a conscious subject. In any case even if 'failing to conform' were so construed we still could not distinguish between the two cases, for on this construction neither of our two unconscious drivers would be guilty. Similarly, if we were to regard the question of responsibility as depending, not on the meaning of the words 'fails to conform' but on the general doctrine that even in cases of strict responsibility the accused must, if he is to be convicted, have been conscious when he failed to do the action which the law requires, then both the unconscious drivers would have to be acquitted. Perhaps because of the way in which they have approached such cases, the courts will find themselves unable to draw the distinction in cases of shooting the lights which they have succeeded in drawing in the case of dangerous driving. This would be a pity, but it is true that, in order to distinguish between our two unconscious drivers who shoot the lights, we should have to abandon the idea that 'fails to conform' to a traffic sign requires, *either* as a matter of the meaning of the words *or* as a matter of general doctrine, a subject who was conscious and able to control his movements at the time he commits this offence. Instead we should have to interpret the statement that liability for this offence is 'strict' or that it is an offence of 'absolute prohibition' to mean that if the accused has lost conscious control over his movements it is necessary and also sufficient for liability that, by the exercise of reasonable care, he could have prevented his loss of conscious control over his movements resulting in a breach of the law. If it is argued that this blurs the line between 'strict liability' and negligence, and brings in the latter in a 'subjective' form, I would admit the charge. But I would urge that we do not know how strict 'strict' liability really is, or how absolute 'absolute' prohibition really is, or how 'subjective' negligence is, till we see what the courts do with these ideas in practice.

V

INTENTION AND PUNISHMENT

I

IT is occasionally profitable to approach the philosophy of punishment by a more empirical route than is usually followed. We should, I think, lay aside temporarily—but only temporarily—our concern with those large-scale general attitudes to the morality of punishment which are called theories of punishment, and which bear such labels as 'retributive', 'utilitarian', 'deterrent', 'reformative', and the like. Instead, we should for a time attend more closely to certain prominent features of the punishment of crime which are common to the legal systems of most reputedly civilized countries, and ask, in relation to them, such questions as the following:

(i) Why should the law define the offences which it punishes in such a way as to make this state of mind or will, and not that, a necessary condition of liability to punishment?

(ii) Why should this kind of behaviour be more severely punished than the same kind of behaviour if it is accompanied by this state of mind or will rather than that?

The general virtue of this kind of approach is that it thrusts upon our attention a great variety of situations to which in practice the principles of punishment have to be applied, and this variety is both much greater and much more realistic than any we could think up for ourselves. Hence, problems are disclosed which otherwise might have escaped our attention and gone altogether uncategorized by our general theories of punishment. But most important of all, this approach, as I hope to demonstrate here, will force us to refine the deceptively simple-looking ideas of retribution, or deterrence, and other general notions in terms of which our general theories are framed. For if we are to use our general theories to explain or justify our practice we shall be forced to split them into variants and sub-types and perhaps, indeed, to mould them into new theories, bearing only a tenuous relationship to the old. Of course, it may well be that if our theories do not fit the uniform practices of legal systems we should, at the end, say 'So much the worse for the law. It is, as we have all along suspected, an incoherent, as well as a barbarous and not very effective, business.' On the other hand, we might say 'So much the worse for our theories, if they cannot account for or accommodate long-established and uniform practices and distinctions.' But whichever we do, we shall, I think, after this type of enquiry, command a much better view of the total situation, be better able to avoid the dangers of over-simplification and better placed to make a final choice of our favourite theory of punishment, before clutching it to our breast.

In this essay I shall try out in a limited field this more empirical approach by focusing attention on the place which the criminal law of most countries allocates to the idea of intention, as one of the principal determinants both of liability to punishment and of its severity. All civilized penal systems make liability to punishment for at any rate serious crime dependent not merely on the fact that the person to be punished has done the outward act of a crime, but on his having done it in a certain state of frame of mind or will. These mental or intellectual elements are many and various and are collected together in the terminology of English jurists under the simple sounding description of *mens rea*, a guilty mind. But the most prominent, of these mental elements, and in many ways the most important, is a man's intention, and in English law and in most other legal systems intention, *or something like it*, is relevant at two different points. It is relevant first at the stage before conviction when the question is 'Can this man be convicted of this crime?'—even if, in fact, he will not actually be punished. At this stage, so far as *mens rea* is concerned, it is normally, though not quite always, sufficient, and normally, though not quite always, necessary that the accused did the particular act forbidden by law, and did it intentionally or with something like intention. It is true that it is not always sufficient, because sometimes duress or provocation or certain forms of mental abnormality may become relevant, and in such cases the accused may not be convicted for a particular crime even if he intended to do the act forbidden by law. But the scope of these matters is small indeed; provocation, for example, is in English law limited to homicide, and duress does not extend to it. So intention, or something like it, is usually, though not quite always, *sufficient* so far as murder is concerned, for conviction of a man who has killed another. Intention to do the act forbidden by law, or something like it, is also generally *necessary* for serious crime, though there are exceptions. Gross unthinking negligence may be enough, e.g. in certain cases of manslaughter, as is carelessness in certain motoring offences, and there are also certain forms of strict liability when a man may be liable for punishment even though he did not intend to do what the law forbids, and was not even guilty of negligence. In regard to murder, the law, in certain types of case, has applied doctrines of 'constructive' murder and 'objective' tests of liability so as to render the question of the accused's actual intention largely irrelevant. But these are exceptions, and there is no doubt of the central importance of intention or something like it when the question is, 'Is the accused liable for punishment?'

Intention is also relevant when the stage of conviction is past, and the question is, 'How severely is the accused to be punished?' This is the stage of sentencing, as distinguished from conviction. Sometimes the legislature will mark off a greater maximum penalty for things done with a certain intention than for the same thing done without that intention. So wounding with intention to kill, (even though the victim is not killed) or wounding with intention to resist arrest, is punishable with the maximum penalty of life imprisonment,[1] and thus much more severely than the simple offence of 'unlawfully and maliciously wounding' a man, for which the

[1] Offences against the Person Act. 1861, s.18.

maximum penalty is five years.[2] Sometimes, however, the greater severity of punishment is settled not by the legislature, but is a matter for the judge to settle within the exercise of his discretion, and in doing this he may often allow the question of intention to weigh. However, too much importance should not be attached to these varying maximum penalties in the case of statutory crimes, since many different features may account for the variation. No philosophical principles, presumably, are needed or competent to account for the fact that, whereas maliciously damaging a work of art is punishable with a maximum of six months imprisonment, maliciously damaging textiles is punishable with imprisonment for life.[3]

2

The law's concern with intention at these two different stages (conviction and sentence) generates a number of problems, some of which I consider here. But there is first an analytical question to be faced before we reach these problems. What, after all, is a man's intention? This is a question which, quite apart from the law, philosophers have found both intriguing and enormously difficult to answer in any simple terms, and there is an additional difficulty in the case of the law. For though jurists or expositors of the criminal law will certainly speak as I have done up to now, and say that the notion of a man's intention is relevant to his criminal responsibility at many important points, what they refer to is the use by the law of a concept which, though it corresponds at many points to what is ordinarily meant in non-legal use by intention, cannot be said to be identical with it. It is for this reason that I have used above the more guarded expression 'intention or something like it'. The concept which legal theorists speak of

[2] Ibid., s.20.

[3] Malicious Damage Act. 1861, ss. 39 and 14.

and define as intention diverges from its counterpart in ordinary use at certain points which are of immediate interest to the philosophy of punishment.

The issue is complicated by two further points. In the formal definition of offences to be found in statutes, the word 'intentionally' is extremely rare; on the other hand, the words 'maliciously' or 'wilfully' are frequently found and are said by jurists and sometimes judges expounding the law to be equivalent in meaning to 'intentionally', or at least to include what that word means. But it is also the case that jurists who make use of the words 'intention' and 'intentionally' in expounding the law differ as to its meaning. Some consider that so far as these words signify that a man foresaw or believed that his conduct would have certain consequences they are properly used only where he believed that these would *certainly* occur, and that where their occurrence was merely thought likely to occur the appropriate description is not in terms of intention but of recklessness. Other legal writers extend the use of the word 'intentionally' to cases where the consequences are thought likely, and reserve 'recklessly' for the cases where a man does not assess the consequences as likely. But this semantic dispute is often very barren, since those who insist on the narrower use of intention agree, even if regretfully, with the advocates of the wider use, that under the existing law it is usually enough for criminal liability that the consequences were thought likely, so that the distinction they draw between 'intentionally' and 'recklessly' is at present in most cases immaterial.

What in the law corresponds to intention can, notwithstanding these difficulties, be made clear in the following way. Intention is to be divided into three related parts, to which I shall give what I hope are three self-explanatory names. The first I shall call 'intentionally doing something'; the second 'doing something with a further intention', and the third 'bare intention' because it is the case of intending to do something in the future without doing anything to execute this intention now. The following are simple legal examples of these three aspects of intention. Suppose first that a man has done something which fits the definition of a crime so far as concerns the outward movements of his body which he has made and the harmful consequences; he has, for example, fired a gun at and thereby wounded or killed another man. On these facts the question then arises, 'Did he wound (or kill) him intentionally?' Though these facts may, in the absence of further evidence, entitle others to conclude (both in and out of the law) that the answer to this question is 'Yes', further evidence may show that he wounded or killed the other accidentally owing to a mistaken belief that the gun was unloaded, and so did it unintentionally. With such cases contrast those illustrating the second aspect of intention: doing something with a further intention. A man gets into a dwelling house at night and the question is not, or not merely, 'Did he do that intentionally?' but 'Did he do that with the further intention, or (as lawyers like to say,) "with the intent", of stealing something?' If so he is guilty of burglary, even if in fact he did not steal anything. Many statutory crimes are framed in terms of such further intent, including the crime already mentioned of wounding with intent to kill, or with an intent to resist arrest, as contrasted with a simple 'malicious' wounding which, subject to the considerations mentioned below, is equivalent merely to intentionally wounding. The third aspect of intention, bare intention, or just intending to do something in the future without taking any present steps towards its execution, is not, for reasons which I shall mention later, of central importance in the criminal law, though it is important in the civil law. Indeed, a landlord's right to eject a tenant on the termination of a lease may depend on the question whether he intended before its expiration to reconstruct the premises. Of these three notions the first, intentionally doing something, is for legal purposes the most important, and I shall begin my discussion of problems by pointing to the divergence between the legal use of this notion and common usage. This shows itself in the following way.

If we consider the names and definitions of various crimes (e.g. murder, assault, malicious wounding, etc.) we can see that, in addition to the mental or volitional elements involved, these definitions generally include three distinguishable components:

(i) a movement or movements of parts of the body made by the agent in a certain physical environment;
(ii) the consequences or upshot, usually of a harmful kind, resulting from these movements; and
(iii) a reference to some special setting of circumstances.

For the moment we may neglect the latter, and focus on the first two. Thus, when a man kills another by shooting, he makes certain movements with his finger to pull the trigger of the gun and this, if it is loaded, has as its consequences or upshot the death of the person killed. Consider in this way the facts of a famous Victorian case, *R.* v. *Desmond, Barrett and Others*.[4] In 1868 there lay in jail two Irish Fenians, whom the accused attempted to liberate. For the purpose, one of them, Barrett, dynamited the prison wall outside the area where he mistakenly believed they would be at exercise. Though the plot failed, the explosion killed some persons living nearby. In this case we can distinguish Barrett's movements made in igniting the fuse from the harmful upshot or consequences in the death of the victims Given such facts, we may ask outside the law, 'Did he kill those men intentionally?' Inside the law the cognate question is 'Did he kill them with such malice aforethought as is required to constitute murder?' Generally speaking, so far as any question like that of intentionally harming is concerned, the law, though it may also be content with less, is content to hold a man guilty if the harmful consequence, e.g. death, was foreseen by the accused in the sense that he believed that it would come about as a result of some voluntary action on his part. Whether he thought this would be certain or only likely to ensue may, as I have said, determine the choice between the words 'intentionally' or 'recklessly' for the description of the case, but will not affect the accused's liability. But the point to be observed here is that, for the law, a foreseen outcome is enough, even if it was unwanted by the agent, even if he thought of it as an undesirable by-product of his activities, and in Desmond's case this is what

[4] *The Times*, 28 April 1868 (hereinafter referred to as "Desmond's case").

the death of those killed by the explosion was. It was no part of Barrett's purpose or aim to kill or injure anyone; the victims' deaths were not a means to his end; to bring them about was not his reason or part of his reason for igniting the fuse, but he was convicted on the ground that he foresaw their death or serious injury. As Lord Coleridge said in Desmond's case, it is murder 'if a man did [an] act not with the purpose of taking life but with the knowledge or belief that life was likely to be sacrificed by it.'

The law therefore does not require in such cases that the outcome should have been something intended in the sense that the accused set out to achieve it, either as a means or an end, and here the law diverges from what is ordinarily meant by expressions like 'he intentionally killed those men'. For outside the law a merely foreseen, though unwanted, outcome is not usually considered as intended, and this is so in big matters as well as small. The neighbour who for the pleasure of the music plays her gramophone at 6 a.m., well knowing from my frequent complaints that it will wake me from sleep, as it does, would not normally be said to have intentionally woken me up or woken me up intentionally, any more than Barrett in ordinary parlance would have been said to have killed the victims of the explosion intentionally. The exceptions to this usage of 'intentionally' are cases where a foreseen outcome is so immediately and invariably connected with the action done that the suggestion that the action might not have that outcome would by ordinary standards be regarded as absurd, or such as only a mentally abnormal person would seriously entertain: the connexion between action and outcome seems therefore to be not merely contingent but rather to be conceptual. Thus if a man struck a glass violently with a hammer, knowing that the blow would break it, he would be said to have broken the glass intentionally (though not, perhaps, to have intentionally broken the glass), even if he merely wanted the noise of the hammer making contact with the glass to attract attention. Some legal theorists, Bentham among them, have recorded this divergence by distinguishing (as '*oblique* intention'), mere foresight of consequences from '*direct* intention where the consequences must have been contemplated by the accused not merely as a foreseen outcome but as an end which he set out to achieve, or as a means to and end, and constituted at least part of his reason for doing what he did.

Of course, in such cases of oblique intention where the harmful consequences are merely a foreseen but unwanted by-product of action we would not say that the agent did what he did *un*intentionally: it would be most misleading to say that Barrett unintentionally killed the victims, if he realized that they would be killed, or that my neighbour woke me up unintentionally, because that would suggest that they had not even foreseen the outcome, and had brought it about accidentally or through some mistake. In fact, it is impossible to squeeze such cases into the dichotomy of intentionally and unintentionally, and we must turn to some other expression like 'knowingly' to characterize them. Needless to say, though the law does not *require* in such cases direct intention, it is satisfied with it. A man, believing himself to be a hopeless shot, who shoots in order to kill and succeeds contrary to his expectations is guilty of murder.

It is perhaps easy to understand why when a man is accused of killing and the question is 'Is he to be convicted and so liable to punishment for this?', as distinguished from 'How severely shall we punish him?', the law should neglect the difference between oblique and direct intention, and why lawyers should come as they have in England to use the beautifully ambiguous expression, 'He contemplated this outcome', to cover both. The reason is, I suggest, that both the case of direct intention and that of oblique intention share one feature which any system of assigning responsibility for conduct must always regard as of crucial importance. This can be seen if we compare the actual facts of the Desmond case with a case of direct intention. Suppose Barrett shot the prison guard in order to obtain from them the keys to release the prisoners. Both in the actual Desmond case and in this imaginary variant, so far as Barrett had control over the alternative between the victims' dying or living, his choice tipped the balance; in both cases he had control over and may be considered to have chosen the outcome, since he consciously opted for the course leading to the victims' deaths. Whether he sought to achieve this as an end or a means to his end, or merely foresaw it as an unwelcome consequence of his intervention, is irrelevant at the stage of conviction where the question of control is crucial. However, when it comes to the question of sentence and the determination of the severity of punishment it may be (though I am not at all sure that this is in fact the case) that on both a retributive and a utilitarian theory of punishment the distinction between direct and oblique intention is relevant.

3

Before I consider this last question further, let me turn to a system which does recognize the distinction between direct and oblique intention where English law does not. In Catholic moral theology the so-called doctrine of 'double effect' is used to draw distinctions between cases in a way which is certainly puzzling to me and to many other secular moralists. This doctrine has its most interesting application where doctors may consider taking steps which will accelerate a patient's death. The simplest case is that of the administration of drugs to relieve the pain of a person slowly dying in agony. According to the latest Papal pronouncements a distinction must be drawn between the case where the drug is given and the patient ceases to feel pain, but as a further consequence his death is accelerated, and the case where he ceases to feel pain because a drug has been administered to kill him as the only way of saving further pain. In the first case, the acceleration of death and the extinction of pain are both effects of the drug, but independent of each other; in the second case, the extinction of pain is not causally independent of the death, and the death is not merely a foreseen but unwanted outcome, but is sought as a means to the extinction of pain. The causal connexion runs *through* the death to the extinction of pain. The use of the drug is forbidden as an instance of direct killing in the second case, but permitted in the first case. The same doctrine would forbid what was done in the case of which I have been told by an eye-witness, where a man had been trapped in the cabin of a blazing lorry from which it was impossible to free him, and a bystander, in answer to his pleas, shot him and killed him to save him from further agony as he was slowly being burnt to death.

The doctrine of double effect is said to distinguish those cases where, according to Catholic doctrine, a doctor may save the life of a pregnant woman at the cost of the life of the foetus from those where he is not permitted to do this. But the contrasting examples usually cited seem to me either not to illustrate this doctrine but some other way of drawing a distinction between killing (or 'direct killing') and an act or omission having death for its consequence or to depend rather implausibly on the point mentioned above, that in certain cases a foreseen but unwanted outcome will be taken to be intended if it is of a kind so immediately and invariably connected with action of the kind done that the connexion is regarded as conceptual rather than contingent. Thus, if a woman is found to have cancer of the womb of which she will die unless the womb is removed, the surgeon may, according to Catholic doctrine, remove the womb with the foreseen consequence that the foetus dies. On the other hand, he is not permitted to perform a craniotomy killing an unborn child to save a woman in labour who would die if the head of the foetus is not crushed. Yet in such cases it could be argued that it is not the death of the foetus but its removal from the body of the mother which is required to save her life; in both cases alike the death of the foetus is a 'second effect', foreseen but not used as a means to an end, or an end. Hence, if the craniotomy is contrasted with the removal of the womb containing the foetus as a case of 'direct' killing, it must be on the basis that the death of the foetus is not merely contingently connected with the craniotomy as it is with the removal of the womb containing it. But it is not clear that the supposition of the survival of the foetus makes better sense in the one case than the other. There are however quite clear cases which illustrate the distinction between direct and indirect intention. Thus, if a doctor found it necessary to kill the foetus while still attached to the wall of the womb by altering the chemical composition of the amniotic fluid with a saline solution in order to avoid the risks of surgery, this would be a clear case of direct intention (since the death of the foetus would in this case be a means to an end) to be contrasted with both the two cases mentioned above, which are arguably both cases of double effect.

Perhaps the most perplexing feature of these cases is that the overriding aim in all of them is the same good result, namely in the first group to save human suffering and in the second to save the mother's life. The differences between the cases are differences of causal structure leading to the applicability of different verbal distinctions. There seems to be no relevant moral difference between them on any theory of morality. It is perfectly true that those cases that the Catholic doctrine forbids may be correctly described as cases of intentional killing (intentionally killing the dying man to stop his pain, intentionally killing the unborn child to save the mother), whereas the cases which the doctrine allows are naturally described as cases of 'knowingly causing death'. But neither these verbal differences nor the differences in causal structure are correlated with moral factors. In certain cases there may be some concomitant moral differences, but these seem to be only contingently connected with the difference between direct and indirect intention. Thus in the Desmond case—one of indirect intention—there was logically room for Barrett to hope, as no doubt he did, that no one would be killed, although he realized that this was most unlikely. But one could easily construct examples to exclude even this factor. The doctor removing, as he may according to the Catholic doctrine, a cancerous womb cannot hope that the foetus which it contains will survive. He may regret its death, but so he may, and no doubt does, regret the death in the case of direct intention, where he kills the unborn child to save the mother's life. It seems that the use of the distinction between direct and oblique intention to draw the line, as Catholic doctrine does, between what is sin and what is not sin, in cases where the ultimate purpose is the same, can only be explained as the result of a legalistic conception of morality as if it were couched in the form of a law in rigid form prohibiting all intentional killing as distinct from knowingly causing death.

4

The English law of abortion, in common with other secular systems, makes no distinction between the two ways of terminating a pregnancy which I have illustrated above: if undertaken to save the mother's life, both forms of destroying the foetus are permissible, and the distinction between direct and indirect intention is ignored as usual, at this stage. But the law courts do apparently recognize this distinction between oblique and direct intention when they are confronted with that second aspect of intention which I have called further intention: that is, when the offence is defined as doing one thing with the intention that something else shall occur or be done, like wounding with intent to kill, or doing an act likely to assist the enemy with the intention of assisting the enemy. Here it seems that it is not enough that the accused believed that there would be a certain outcome; for conviction on such a charge it must be shown that he contemplated that outcome as an end or as a means to some end. The case most frequently cited in support of this view of the law is *R.* v. *Steane.*[5] The facts of this case were that Steane, who was resident in Germany on the outbreak of war, was threatened by the Gestapo that he and his wife would be beaten up unless he broadcast enemy propaganda in English. To save them, he did broadcast, and after the war was charged with the offence of doing an act likely to assist the enemy with the intention of assisting the enemy. He was, in fact, acquitted on the ground that unless it was shown that he broadcast in

[5] (1947) K.B. 997

order to assist the enemy the charge was not made out, and his mere knowledge that it was likely to assist the enemy was not enough. The decision seems to me to be based on a correct interpretation of the statutory language; but the moral or policy justification for acquitting Steane on this ground has seemed far from clear to many critics. Some of them have urged that, though it might have been perfectly sensible to acquit Steane on the ground of duress, the ground on which he was acquitted, adhering very strictly as it does to the meaning of the words used in defining the offence, might have very odd consequences. Thus, if it was necessary in order to satisfy the language of the statute to show that the accused broadcast in order to assist the enemy, as distinct from merely knowing or believing that his broadcasts would assist them, it would seem that he should also have been acquitted if instead of being threatened with beating he was promised a packet of cigarettes and broadcast in order to get them, the assistance given to the enemy in both cases being a foreseen but unwanted consequence of broadcasting undertaken for other reasons.

But this same distinction is, according to most authorities, also made by English law in dealing with the notion of an attempt to commit a crime. For this notion, too, involves the idea of doing something with the direct intention that some consequence should come about, as distinct from merely doing it with the belief that it would come about. A hypothetical case has been used by Professor Glanville Williams to illustrate the absurdity that the application of this distinction might have in the case of attempted murder. Suppose one man is walking with another along the edge of a cliff and sees a diamond ring on the path before him. Knowing that his companion also wishes to get the ring, he pushes him over the cliff, believing that this will in all probability lead to his death, but, in fact, a bush breaks his fall a short distance down the cliff, and he is unharmed. This, according to the usual interpretation of the notion of an attempt, probably does not constitute an attempt to murder, for A did not push his companion over in order to kill him, though he believed it would cause his death; whereas if, in order to get the ring, instead of pushing the victim over the cliff, A had shot at him to kill him but missed, this would have been attempted murder. Yet the ultimate aim is the same in both cases.

Can one formulate any intelligible theory of punishment which would make sense of this distinction? No calculation of the efficacy of deterrence or reforming measures, and nothing that would ordinarily be called retribution, seems to justify this distinction. In the attempt case, for example, the variant where the direct intention to kill is present and the variant where the intention is indirect seem equally wicked, equally harmful, and equally in need of discouragement by the law. The distinction seems to make its appeal to a feeling that to *use* a man's death as a means to some further end is a defilement of the agent: his will is thus identified with an evil aim and it is somehow morally worse than the will of one who in the pursuit of the same further end does something which, as the agent realizes, renders the man's death inevitable as a second effect. If this is the basis of the distinction we may well ask whether the law should in such cases give recognition to it, especially where, as in the attempt case, recognition will lead to an acquittal, except on a relatively trivial charge of assault.

I shall leave the topic of direct and oblique intention to consider the more general problems generated by the law relating to attempts. As everyone knows, a bare intention to commit a crime is not punishable by English law. This has been often repeated from the Bench since Lord Mansfield in 1784 said: 'So long as an act rests in bare intention alone it is not punishable by our law.' The reasons for this are perhaps not far to seek. Not only would it be a matter of extreme difficulty to ferret out those who were guilty of harbouring, but not executing, mere intentions to commit crimes, but the effort to do so would involve vast incursions into individual privacy and liberty. The Victorian judge, James Fitzjames Stephen, said that to punish bare intention 'would be utterly intolerable: all mankind would be criminals, and most of their lives would be passed in trying and punishing each other for offences which could never be proved.'[6] But the law, though

[6] Stephen, *A History of the Criminal Law of England*, Vol. II, p. 78.

it does not punish bare intention, does punish as an attempt the doing of something quite harmless in itself, if it is done with the further intention of committing a crime and if the relationship between the act done and the crime is sufficiently 'proximate' or close. So the would-be thief who puts his hand into a pocket which contrary to his expectations proves empty, or who writes a letter to obtain money by false pretences which fails to deceive his intended victim, or the would-be murderer who puts poison into the cup which is emptied before the intended victim can drink out of it, are all guilty of attempts to commit crimes.

It is not obvious, however, at least on some versions of utilitarian theory of punishment, why attempts should be punishable, as they are, in most legal systems. On a retributive view perhaps the answer is easy. The criminal had gone so far as to do his best to execute a wicked intention, and the difficulties of proof and so on are removed by his overt act. But on what is generally known as a deterrent theory the case for punishing attempts has seemed, even to some of its supporters, unclear. Thus it has been argued that if we think of the law, as the deterrent theory requires us to think of it, as threatening punishment to those persons who are tempted to commit offences, there can be no need to attach any punishment to the unsuccessful attempt, because those who set about crime intend to succeed and the law's threat has all the deterrent force it can have if it is attached to the crime; no additional effect is given to it if unsuccessful attempts are also punished. This is a fallacy, but I shall dwell a little on the point because it shows us how careful we must be to distinguish between various aspects of the idea of a deterrent theory of punishment. First we must distinguish between what is called the general deterrent, consisting of the threat of punishment to all who are tempted to commit offences, and the individual deterrent, consisting not merely of the threat of punishment for future offences, but also of the application of punishment to individuals who have not been deterred by the law's threats and have actually committed the offence. If we make this distinction there seems a clear case for the use of punishment as an individual deterrent in the cases of unsuccessful attempts to commit crimes; for the accused has manifested a dangerous disposition to do all he can to commit a crime, and the experience of punishment may check him in the future, since it may cause him to attach more weight to the law's threats. From this point of view the punishment of a man who has attempted but failed seems as well justified on deterrent grounds as the punishment of a man who has succeeded in committing a crime, though I shall consider later the usual practice of punishing the attempt less severely. But even from the point of view of the general deterrent, the sceptical argument which suggested that there is no case for punishing an attempt is, after all, mistaken. It is perfectly true that those who commit crimes intend to succeed, but this does not show that punishing a man for an unsuccessful attempt will not increase the efficacy of the law's threats, or that failure to punish him would not often diminish their efficacy. This is so for two reasons: first, there must be many who are not completely confident that they will succeed in their criminal objective, but will be prepared to run the risk of punishment if they can be assured that they have to pay nothing for attempts which fail; whereas if unsuccessful attempts were also punished the price might appear to them to be too high. Again, there must be many cases where men might with good or bad reason believe that if they succeed in committing some crime they will escape, but if they fail they may be caught. Treason is only the most obvious of such cases, and unless attempts were punished, there would, in such cases, be no deterrent force in the law's threat attached to the main crime.

A more difficult question concerns the almost universal practice of legal systems of fixing a more severe punishment for the completed crime than for the mere attempt. How is this to be justified? Here a retributive theory in which severity of punishment is proportioned to the allegedly evil intentions of the criminal is in grave difficulty; for there seems to be no difference in wickedness, though there may be in skill, between the successful and the unsuccessful attempt in this respect.

Very often an unsuccessful attempt is merely the accidental failure to commit the crime because somebody unexpectedly intervenes and frustrates the attempt. As far as I can see a deterrent theory, except in relation to a very specialized class of crimes, is in similar difficulties. The exceptions, which have been mooted since Beccaria first discussed them, are those crimes whose consummation occupies a considerable space of time so that the criminal may have time between the attempt and its consummation to think again. He may have what is called a *locus poententiae*, and he might desist, but if he is already involved in the full penalty by virtue of merely having attempted to commit the crime, he may have no motive for desisting. Similar reasoning is presented when it is pointed out that if a man shoots and misses there is no reason why he should not shoot again if he is already liable to the full penalty for his unsuccessful attempt. Such cases are, of course, realities; but they are surely very rare, if only because in the law of most systems in order to be guilty of an attempt one has to get very near to the completion of the full offence, and the question of a second shot may arise only seldom. Yet apart from this, there seems no reason on any form of deterrent theory, whether we consider the general deterrent or the individual deterrent, for punishing the unsuccessful attempt less severely than the completed crime. The individual who has tried but failed to carry out the planned crime may need just as much punishment to keep him straight in the future as the successful criminal. He may be as much disposed to repeat his crime.

The almost universal tendency in punishing to discriminate between attempts and completed crimes rests, I think, on a version of the retributive theory which has permeated certain branches of English law, and yet has on occasion been stigmatized even by English judges as illogical. This is the simple theory that it is a perfectly legitimate ground to grade punishments according to the amount of harm actually done, whether this was intended or not; 'if he has done the harm he must pay for it, but if he has not done it he should pay less'. To many people such a theory of punishment seems to confuse punishment with compensation, the amount of which should indeed be fixed in relation to harm done. Even if punishment and compensation were not distinguished in primitive law, many think that this is no excuse for confusing them now. Why should the accidental fact that an intended harmful outcome has not occurred be a ground for punishing less a criminal who may be equally dangerous and equally wicked? I may be wrong in thinking that there is so little to be said for this form of retributive theory. It is certainly popular, and the nearest to a rational defence that I know of it is the following. It is pointed out that in some cases the successful completion of a crime may be a source of gratification, and, in the case of theft, of actual gain, and in such cases to punish the successful criminal more severely may be one way of depriving him of these illicit satisfactions which the unsuccessful have never had. This argument, which certainly has some attraction where the successful criminal has hidden loot to enjoy on emerging from prison, would be an interesting addition to theories of punishment of the principle that the wicked should not be allowed to profit by their crimes.

My own belief is that this form of retributive theory appeals to something with deeper instinctive roots than the last mentioned principle. Certainly the resentment felt by a victim actually injured is normally much greater than that felt by the intended victim who has escaped harm because an attempted crime has failed. Bishop Butler, in his sermon on resentment explains on this ground the distinction men draw between 'an injury done' and one 'which, though designed, was prevented, in cases where the guilt is perhaps the same'. But again the question arises, if this form of retributive theory depends on the connexion between blame and resentment, whether the law should give effect to such a theory. Can we not control resentment, however natural, in the interests of some deliberate forward-looking policy, much as we control our natural fears in the interest of forward-looking prudential aims? And if we can do this, should we not do so? And might not this require us in some cases to punish attempts as severely as the completed offence?

Now to my last topic. I have said that the intention, as the law understands it (that is either oblique or direct intention), is generally, though not always sufficient and generally necessary for criminal liability. But it is not always so, and in conclusion I shall consider certain cases where a man may be punished for a crime, although he had no intention, oblique or direct, to do the act forbidden by law. Here too, we shall be forced in the end, if we are to fit any theories of punishment to the facts, to ask what we mean precisely by the idea of retribution or deterrence, and to refine these ideas in further new directions.

There are two main types of case where no form of intention is required for criminal liability. The first type consists of those crimes known as crimes of strict liability, where it is no defence to prove that you did not intend to do the act forbidden, did not know that you were doing it, and indeed took every care to avoid doing it. Strict liability of this sort is usually thought odious by academic writers, even though most of the offences in question are punished with fairly minor penalties. They include such things as selling liquor to an intoxicated person, or selling adulterated milk, or driving a car without insurance; though they also include more serious offences. The second type of case on which I shall spend a little time are cases of negligence, and I shall concentrate on this, since some useful lessons can be learnt from it.

English law is rather sparing in its punishment of negligence in the sense of unintentional neglect to take reasonable precautions against harm to others, and apart from a few isolated situations, it can be said that negligence is only punishable as a crime if it results in death (where it constitutes one species of manslaughter) or if it is shown in driving motor vehicles on the road: not only is there the offence of driving without due care and attention but there is also the more severely punished offence of causing death by dangerous driving. There are no doubt some general historical reasons why the law should be slow to punish a man who is negligent in the sense of not realizing, though he ought to have realized, that his actions or omissions might occasion a serious harm; but in addition to these historical grounds it is the case that many lawyers feel uncomfortable in accepting negligence as a basis of criminal liability, though, in fact, in our ordinary life it is not usually held to be a good excuse to say 'I didn't mean to do it' or 'I just didn't stop to think'. Certain jurists draw a very dramatic line between the cases where a man intended or foresaw that his actions would be harmful, and those where he was grossly careless, but, as they say, 'inadvertent'. To such lawyers it appears that the 'state of mind' of the merely careless man is not in any way wicked, and presents nothing on which a retributive theory can, as it were, bite. But what is more extraordinary is that very many utilitarian-minded jurists think that there is no intelligible case for punishing gross inadvertent negligence as a deterrent. This seems to conflict very much with the common-sense belief that in some cases we may make people more careful by blaming or punishing them for carelessness. But the scepticism has ancient, if not respectable, roots. In fact, it derives from a general conception of the notion of deterrence which you will find explicit in Bentham and his follower, Austin, who gives a rather restricted interpretation to the notion that a deterrent punishment must work 'through the mind' of those who are to be deterred. These thinkers conceived the law's threats of punishment as something which would enter into the reasoning and deliberation of the potential criminal at the moment when he considered whether or not to commit the crime: the threats were to constitute, for the person tempted to commit the crime, reasons against committing it, and the hope was that the reasons would appear conclusive and lead to a decision to conform. In this rationalistic picture of what one might call 'criminal deliberations', the threat of punishment was intended to constitute a *guide* to deliberation on the assumption that he would be tempted to commit the crime and he would deliberate.

Now, it is plain that on this conception the threat of punishment could not be a guide to those who committed their crime through inadvertent unthinking negligence, for *ex hypothesi* in such cases there was no moment in which they were tempted and deliberated whether to commit the crime or not. And, in fact, many writers, including Professor Glanville Williams, have shared an assumption that to be a deterrent the threat of punishment must be capable of entering into the deliberations of the criminal as a guide at the moment when he contemplates his crime. But surely this is an excessively narrow way of conceiving the relevance of threats to conduct. Threats may not only guide your deliberations—your practical thinking—but may cause you to think. A man drives a car with one arm round his girl friend's neck and gazes into her eyes instead of at the road, and is subsequently punished for careless driving in spite of his protestations that he 'just didn't think' of the possibilities of the harmful outcome to others on the road. Surely it is not absurd to hope that, as a result, next time he drives he may approach his car, and perhaps also his girl friend, in a very different spirit. Recollection of the punishment and the knowledge that others are punished may make a driver think; and if he thinks (since he has no intention to drive badly) he may say to himself 'this time I must attend to my driving', and the effect may well be that he drives with due care and attention. No doubt the connexion between the threat of punishment and subsequent good behaviour is not of the rationalistic kind pictured in the guiding-type of case. The threat of punishment is something which causes him to exert his faculties, rather than something which enters as a reason for conforming to the law when he is deliberating whether to break it or not. It is perhaps more like a goad than a guide. But there seems to me to be nothing disreputable in allowing the law to function in this way, and it is arguable that it functions in this way rather than in the rationalistic way more frequently than is generally allowed. At any rate, consideration of the punishment of negligence (and also punishment in the strict liability cases) brings out the need to refine in this way the idea of deterrence by threat.

The punishment of negligence has in England, and I suspect in most legal systems, some further curious features, for here the severity of punishment is often determined by the seriousness of the outcome, and the explanation of this involves recourse again to that sense of retribution which I have already mentioned in explaining the lighter punishments normally accorded to unsuccessful attempts to commit crimes. In 1956, when the offence known as 'causing death by dangerous driving' was created, there was an illuminating debate in the House of Lords, where Lord Hailsham pointed out how absurd it was that, where two people were equally careless in driving on the roads, one of them should be liable to be punished with the severe sentence of five years' imprisonment if the bad driving resulted in someone's death, whereas if it resulted only in the victim being crippled, or if no one was harmed he would be liable only for the maximum penalty of two years'. Of course, a similar 'illogicality', as Lord Hailsham called it, is to be seen in the very existence of one species of manslaughter where the accused's grossly negligent act is punishable if it causes the victim's death. The then Lord Chancellor (Lord Kilmuir) recognized the 'illogicality' of this mode of determining the relevant severity of punishment by reference to the harm done, but said in the debate, 'Such doubts have of course affected every penal thinker and penal reformer. But no one has been able to translate these doubts into a workable system. Results must be taken into account if the penalties are going to have the effects which it is desirable they should have.' I have never understood that answer in defence of this form of retributive theory. Does it merely state obscurely what Stephen stated with great clarity seventy years earlier, 'It gratifies a natural public feeling to choose out for punishment the one who actually has caused great harm.'?[7]

[7] Stephen, op. cit., Vol. III, p. 311.

VI

NEGLIGENCE, *MENS REA* AND CRIMINAL RESPONSIBILITY

'I DIDN'T *mean* to do it: I just didn't think.' 'But you should have thought.' Such an exchange, perhaps over the fragments of a broken vase destroyed by some careless action, is not uncommon; and most people would think that, in ordinary circumstances, such a rejection of 'I didn't think' as an excuse is quite justified. No doubt many of us have our moments of scepticism about both the justice and the efficacy of the whole business of blaming and punishment; but, if we are going in for the business at all, it does not appear unduly harsh, or a sign of archaic or unenlightened conceptions of responsibility, to include gross, unthinking carelessness among the things for which we blame and punish. This does not seem like the 'strict liability' which has acquired such odium among Anglo-American lawyers. There seems a world of difference between punishing people for the harm they unintentionally but carelessly cause, and punishing them for the harm which no exercise of reasonable care on their part could have avoided.

So 'I just didn't think' is not in ordinary life, in ordinary circumstances, an excuse; nonetheless it has its place in the rough assessments which we make, outside the law, of the gravity of different offences which cause the same harm. To break your Ming china, deliberately or intentionally, is worse than to knock it over while waltzing wildly round the room and not thinking of what might get knocked over. Hence, showing that the damage was not intentional, but the upshot of thoughtlessness or carelessness, has its relevance as a mitigating factor affecting the quantum of blame or punishment.

I. THE CRIMINAL LAW

These rough discriminations of ordinary life are worked out with more precision in the criminal law, and most modern writers would agree with the following distinctions and terminology. 'Inadvertent negligence' is to be discriminated not only from deliberately and intentionally doing harm but also from 'recklessness', that is, wittingly flying in the face of a substantial, unjustified risk, or the conscious creation of such a risk. The force of the word 'inadvertent' is to emphasize the exclusion both of intention to do harm and of the appreciation of the risk; most writers after stressing this point, then use 'negligence' simply for inadvertent negligence.[1] Further, within the sphere of inadvertent negligence, different degrees are discriminated: 'gross negligence' is usually said to be required for criminal liability in contrast with something less ('ordinary' or 'civil' negligence) which is enough for civil liability.

In Anglo-American law there are a number of statutory offences in which negligence, in the sense of a failure to take reasonable precautions against harm, unaccompanied either by intention to do harm or an appreciation of the risk of harm, is made punishable. In England, the Road Traffic Act, 1960, affords the best known illustration: under section 10 driving without due care and attention is a summary offence even though no harm ensues. In other jurisdictions, criminal codes often contain quite general provisions penalizing those who 'negligently hurt' or cause bodily harm by negligence.[2] *Pace* one English authority, Dr. Turner (whose views are examined in detail below), the common law as distinct from statute also admits a few crimes,[3] including manslaughter, which can be committed by inadvertent negligence if the negligence is

[1] This terminology is used by Glanville Williams, *Criminal Law, The General Part* (2nd edn.), Ch. III, p. 100 et seq., and also by the American Law Institute Draft Model Penal Code s. 2.0.2 (Tentative Draft 4, p. 26 and Comment, ibid., pp. 126–7). So, too, Cross and Jones, *Introduction to Criminal Law* (5th edn.). pp. 42–45.

[2] See for these and other cases Glanville Williams op. cit., p. 120 n. 22.

[3] Other common law crimes commonly cited are non-repair of a highway and public nuisance. Besides these there are controversial cases including certain forms of murder (*R.* v. *Ward* (1956) 1 Q.B. 351, Cross and Jones, op. cit., pp. 48–52 and *D.P.P.* v. *Smith* (1961), A.C. 290. These cases some writers consider as authorities for the proposition that criminal negligence is sufficient malice for the crime of murder. There are, however, reasons for doubting this interpretation of these cases.

sufficiently 'gross'.[4] It is, however, the case that a number of English and American writers on criminal law feel uneasy about different aspects of negligence. Dr. Glanville Williams[5] thinks that its punishment cannot be justified on either a retributive or a deterrent basis. Professor Jerome Hall[6], who thinks that moral culpability is the basis of criminal responsibility and that punishment should be confined to 'intentional or reckless doing of a morally wrong act', disputes both the efficacy and justice of the punishment of negligence.

In this essay I shall consider a far more thorough-going form of scepticism. It is to be found in Dr. Turner's famous essay *The Mental Element in Crimes at Common Law*[7]. There he makes three claims; first, that negligence has no place in the Common Law as a basis of criminal responsibility, and so none in the law of manslaughter; secondly, that the idea of degrees of negligence and so of gross negligence is nonsensical; thirdly (and most important), that to detach criminal responsibility from what he terms 'foresight of consequences', in order to

[4] See Cross and Jones, op. cit., pp. 152–5. The American Law Institute accepts this view of the English law of manslaughter (Tentative Draft 9, p. 50) but advocates treatment of negligent homicide as an offence of lower degree than manslaughter. Glanville Williams, op. cit., p. 106 (s. 39) after stating that manslaughter can be committed by inadvertent negligence 'for the accused need not have foreseen the likelihood of *death*' says that the 'ordinary formulations' leave in doubt the question whether foresight of some bodily harm (not necessarily serious injury or death) is required for manslaughter. He describes (op. cit., p. 108) as 'not altogether satisfactory' the cases usually taken to establish that no such foresight is required viz. *Burdee* (1916), 86 L.J. K.B. 871, 12 Cr. App. Rep. 153; *Pittwood* (1902), 19 T.L.R. 37; *Benge* (1865), 4 F. & F. 504; *John Jones* (1874), 12 Cox 628. Of *Bateman* (1925), 28 Cox 33; 19 Cr. App. Rep. 8 he says 'it may be questioned whether this does not extend the law of manslaughter too widely' and thinks in spite of *Andrews* v. *D.P.P.* (1937), A.C. 576 that the issue is still open for the House of Lords. (op. cit., pp. 107, 110).

[5] Op. cit., pp. 122–3.

[6] *Principles of Criminal Law*, pp. 366–7, and *43 C.L.R.*, p. 775. Professor Herbert Wechsler (Reporter in the A.L.I. Draft Model Penal Code) rejects this criticism and holds that punishment for conduct which inadvertently creates improper risks 'supplies men with an additional motive to take care before acting, to use their faculties and to draw on their experience in gauging the potentialities of contemplated conduct', Tentative Draft 4, pp. 126–7, and Tentative Draft 9, pp. 52–53.

[7] *The Modern Approach to Criminal Law* (1945), p. 195.

admit negligence as a sufficient basis of such responsibility is necessarily to revert to a system of 'absolute' or strict liability in which no 'subjective element' is required.

Dr. Turner's essay has of course been very influential; he has reaffirmed the substance of its doctrines in his editions of both Kenny[8] and Russell.[9] This, however, is not my reason for submitting his essay to a fresh scrutiny so long after its publication. My reason is that his arguments have a general interest and importance quite independent of his conclusions about the place of negligence in the common law. I shall argue that they rest on a mistaken conception both of the way in which mental or 'subjective' elements are involved in human action, and of the reasons why we attach the great importance which we do to the principle that liability to criminal punishment should be conditional on the presence of a mental element. These misconceptions have not been sufficiently examined: yet they are I think widely shared and much encouraged by our traditional legal ways of talking about the relevance of the mind to responsibility. Dr. Turner's arguments are singularly clear and uncompromising; even if I am right in thinking them mistaken his mistakes are illuminating ones. So much cannot always be said for the truths uttered by other men.

Before we reach the substance of the matter one tiresome question of nomenclature must be got out of the way. This concerns the meaning of the phrase '*mens rea*'. Dr. Turner, as we shall see, confines this expression to a combination of two elements, one of which is the element required if the accused's conduct is to be 'voluntary,' the other is 'foresight' of the consequences of conduct. Dr. Glanville Williams, although he deprecates the imposition of criminal punishment for negligence, does not describe it or (apparently) think of it, as Dr. Turner does, as a form of 'strict' or 'absolute' liability; none-theless, though not including it under the expression 'strict liability', he excludes it from the scope of the term '*mens rea*', which he confines to intention and recklessness. Judicial

[8] Kenny's *Outlines of Criminal Law* (19th edn.), pp. 37–40.

[9] Russell on Crime (12th edn.), pp. 43–44, 52, 62–66.

pronouncements, though no very careful ones, can be cited on either side.[10]

There is, I think, much to be said in mid-twentieth century in favour of extending the notion of '*mens*' beyond the 'cognitive' element of knowledge or foresight, so as to include the capacities and powers of normal persons to think about and control their conduct: I would therefore certainly follow Stephen and others and include negligence in '*mens rea*' because, as I shall argue later, it is essentially a failure to exercise such capacities. But this question of nomenclature is not important so long as it is seen for what it is, and not allowed either to obscure or prejudge the issue of substance. For the substantial issue is not whether negligence should be called '*mens rea*'; the issue is whether it is true that to admit negligence as a basis of criminal responsibility is *eo ipso* to eliminate from the conditions of criminal responsibility the subjective element which, according to modern conceptions of justice, the law should require. Is its admission tantamount to that 'strict' liability which we now generally hold odious and tolerate with reluctance?

2. VOLUNTARY CONDUCT AND FORESIGHT OF CONSEQUENCES

According to Dr. Turner, the subjective element required for responsibility for common law crimes consists of two distinct items specified in the second and third of three general rules which he formulates.

'Rule I—It must be proved that the conduct of the accused person caused the *actus reus*.

Rule II—It must be proved that this conduct was *voluntary*.

Rule III—It must be proved that the accused person *realised*

[10] See Glanville Williams, op. cit., p. 102, n. 8. Examples on each side are Shearman J. in *Allard* v. *Selfridge*, (1925) 1 K.B. 129, at p. 137. ('The true translation of that phrase is criminal intention, or an intention to do the act which is made penal by statute or by the common law') and Fry L. J. in *Lee* v. *Dangar, Grant & Co.*, (1892) 2 Q.B. 337, at p. 350. 'A criminal mind or that negligence which is itself criminal'. See also for a more discursive statement *R.* v. *Bateman* (1925), 19 Cr. App. Rep. 8 *per* Hewart C. J.

at the time that his conduct would, or might *produce results of a certain kind*, in other words that he must have foreseen that certain consequences were likely to follow on his acts or omissions. The extent to which this foresight of the consequences must have extended is fixed by law and differs in the case of each specific crime. . . .'[11]

We shall be mainly concerned with Rule III—as is Dr. Turner's essay. But something must be said about the stipulation in Rule II that the accused's 'conduct' must be 'voluntary'. Dr. Turner himself considers that the truth contained in his Rule III has been obscured because the mental element required to make conduct voluntary has not been discriminated as a separate item in *mens rea*. I, on the other hand, harbour the suspicion that a failure on Dr. Turner's part to explore properly what is involved in the notion of 'voluntary conduct' is responsible for much that seems to me mistaken in his further argument.

Certainly it is not easy to extract either from this essay or from Dr. Turner's editions of Kenny or Russell what is meant by 'conduct', and what the mental element is which makes conduct 'voluntary'. At first sight Dr. Turner's doctrine on this matter looks very like the old simple Austinian[12] theory that what we normally speak of and think of as actions (killing, hitting, etc.) must be divided into two parts (*a*) the 'act' or initiating movement of the actor's body or (in more extreme versions) a muscular contraction, (*b*) the consequences of the 'act'; so that an 'act' is voluntary when and only when it is caused by a 'volition' which is a desire for the movement (or muscular contraction). But such an identification of Dr. Turner's 'conduct' with the Austinian 'act' (or movement of the body), and the mental element which makes it voluntary, with the Austinian volition or desire for movement, is precluded by two things. First, Dr. Turner says conduct includes not only physical acts but omissions. Secondly, though 'conduct' is always something less than the *actus reus* which is its 'result' (e.g. killing in murder) it is by no means confined by him as

[11] *The Modern Approach to Criminal Law* (1945), p. 199.
[12] Austin, *Lectures on Jurisprudence* (5th edn.), Lecture XVIII.

'act' is by Austin to the mere initiating movement of the actor's body. Dr. Turner tells us that 'by definition *conduct*, as such, cannot be criminal'.[13] He also explains that 'conduct is of course itself a series of deeds, each of which is the result of those which have come before it; but at some stage in this series a position of affairs may be brought into existence which the governing power in the state regards as so harmful as to call for repression by the criminal law. It is this point of selection by the law, this designation of an event as an *actus reus*, which for the purposes of our jurisprudence marks the distinction between *conduct* and *deed*.'[14]

About the mental element required to make conduct voluntary, Dr. Turner tells us[15] only that it is a 'mental attitude to [his] conduct' (as distinct from the consequences of conduct) and that if conduct is to be voluntary 'it is essential that the conduct should have been the result of the exercise of the will'. He does however give us examples of involuntary conduct in a list not meant to be exhaustive: 'for example, if *B* holds a weapon and *A*, against *B*'s will, seizes his hand and the weapon, and therewith stabs *C*; and possibly an act done under hypnotic suggestion or when sleep-walking or by pure accident. In certain cases of insanity, infancy and drunkenness the same defence may be successfully raised.'[16]

This account of voluntary conduct presents many difficulties. What is it for conduct to be 'the result of the exercise of the will'? Must the actor desire or will only the initiating movement of his body or the whole course of 'conduct' short of the point when it becomes an *actus reus*? And how does this account of the distinction between the course of conduct and the *actus reus* which is said to be its 'result' apply to omissions? The examples given suggest that Dr. Turner is here grossly hampered by traces of the old psychology of 'act' and 'volition', and no satisfactory account of what it is which makes 'conduct' voluntary or involuntary, capable of covering both acts and

[13] Op. cit., p. 240. [14] Op. cit., p. 239. [15] Kenny (19th edn.), p. 30.

[16] *The Modern Approach to Criminal Law* (1945), p. 204. See the further examples suggested in Kenny (19th edn.), p. 29: viz., when harm 'results from a man's movements in an epileptic seizure, or while suffering from St. Vitus's Dance'.

omissions can be given in his terminology of 'states of mind', or 'mental attitude'. What is required (as a minimum) is the notion of a general *ability* or *capacity* to control bodily movements, which is usually present but may be absent or impaired.

But even if we waive these difficulties, Dr. Turner's twofold account of *mens rea* in terms of 'voluntary conduct' and 'foresight of consequences' is at points plainly inadequate. It does not fit certain quite straightforward, familiar, cases where criminal responsibility is excluded because of the lack of the appropriate subjective element. Thus it does not, as it stands, accommodate the case of mistake; for a mistaken belief sufficient to exclude liability need not necessarily relate to *consequences*; it may relate to *circumstances* in which the action is done, or to the character or identity of the thing or person affected. Of course, Dr. Turner in his edition of Kenny, under the title of 'Mistake as a Defence at Common Law', discusses well-known cases of mistake such as *Levett's case*,[17] where the innocent victim was killed in mistake for a burglar, and says (in a footnote) that the subjective element in such cases relates to the agent's belief in the facts upon which he takes action.[18] He does not think this calls for a modification in his two-limbed general theory of *mens rea*; instead he adopts the view that such mistakes, since they do not relate to consequences, negative an element in the *actus reus* but do not negative *mens rea*. Besides this curious treatment of mistake, there is also the group of defences which Dr. Turner discusses in the same work under the heading of Compulsion,[19] which include marital coercion and duress *per minas*. Here, as the author rightly says, English law is 'both meagre and vague'; nonetheless, confidence in his general definition as an exhaustive account of *mens rea*, has led him into a curious explanation of the theoretical basis of the relevance to responsibility of such matters of coercion or duress. He cites first an example of compulsion in the case of 'a powerful man who, seizing the

[17] (1638), Cro. Car. 538.
[18] Op. cit., (19th edn.), p. 58 n. 3.
[19] Ibid., p. 66

hand of one much weaker than himself and overcoming his resistance by sheer strength, forces the hand to strike someone else' [20] Of this case he says, 'the defence . . . must be that the mental element of volition is absent—the accused, in other words, pleads that his conduct was not voluntary'[21] and to explain this he refers back to the earlier account of voluntary conduct which we have discussed. The author then says that compulsion can take other forms than physical force,[22] and he proceeds to discuss under this head obedience to orders, marital coercion, duress, and necessity. It is, however, clear that such defences as coercion or duress (where they are admitted) lie quite outside the ambit of the definition of voluntary *conduct* given by Dr. Turner: they are not just different instances of *movement* which is not voluntary because, like the case of physical compulsion or that of epilepsy cited earlier, the agent has no control over his bodily movements. Defences like duress or coercion refer not to involuntary *movements*, but, as Austin[23] himself emphasized, to other, quite different ways in which an *action* may fail to be voluntary; here the *action* may not be the outcome of the agent's free choice, though the *movements* of the body are not in any way involuntary.

So far, my objection is that Dr. Turner's formulation of the subjective element in terms of the two elements of voluntary conduct and foresight of consequences leads to a mis-assimilation of different cases; as if the difference between an action under duress and involuntary *conduct* lay merely in the kind of compulsion used. But in fact the definition of *mens rea* in terms of voluntary conduct *plus* foresight of consequences, leads Dr. Turner to great incoherence in the division of the ingredients of a crime between *mens rea* and *actus reus*. Thus in discussing the well-known case of *R.* v. *Prince*[24] (where the accused was found guilty of the statutory offence of taking a girl under 16 out of the possession of her father notwithstanding that he believed on reasonable grounds that she was over 16) Dr.

[20] Ibid. [21] Ibid. [22] Ibid.

[23] Op. cit., p. 417. Notes to Lecture XVIII, 'Voluntary—Double Meaning of the word Voluntary'.

[24] (1875), L.R. 2 C.C.R. 154.

Turner examines the argument that the word 'knowingly' might have been read into the section creating the offence (in which case the offence would not have been committed by the prisoner) and says 'this change would not affect the *mens rea* of the accused person, but it would merely add another necessary fact to the *actus reus*, namely the offender's knowledge of the girl's age'.[25] But there is nothing to support[26] this startling view that where knowledge is required as an ingredient of an offence this may be part of the *actus reus*, not of the *mens rea*, except the author's definition of *mens rea* exclusively in terms of the two elements of 'voluntary conduct' and foresight of consequences'. If knowledge (the constituent *par excellence* of *mens rea*) may be counted as part of the *actus reus*, it seems quite senseless to insist on any distinction at all between the *actus reus* and the *mens rea*, or to develop a doctrine of criminal responsibility in terms of this distinction.

3. NEGLIGENCE AND INADVERTENCE

So far it is plain that, quite apart from its exclusion of negligence, the account of the subjective element required for criminal responsibility in terms of the two elements 'voluntary conduct' and 'foresight of consequences' is, at certain points, inadequate. Dr. Turner's arguments against the inclusion of negligence must now be examined. They are most clearly presented by him in connexion with manslaughter. Of this, Dr. Turner says[27] 'a man, to be guilty of manslaughter, must have had in his mind the idea of bodily harm to someone'. On this view, what is known to English law as 'manslaughter by negligence' is misdescribed by the words; and Dr. Turner

[25] In 'The Mental Element in Crimes at Common Law': *The Modern Approach to Criminal Law* (1945), op. 219.

[26] There is plain authority against it: see *R.* v. *Tolson* (1889), 23 Q.B.D.168 *per* Stephen J. 'The mental element of most crimes is marked by one of the words "maliciously", "fraudulently", "negligently", or "knowingly".'

[27] *The Modern Approach to Criminal Law* (1945), p. 228.

expressly says that judges in trying cases of manslaughter should avoid all reference to 'negligence' and so far as *mens rea* is concerned should direct the jury to two questions:

(i) Whether the accused's conduct was voluntary;

(ii) Whether at the time he either intended to inflict on someone a physical harm, or foresaw the possibility of inflicting a physical harm and took the risk of it.[28]

To treat these cases otherwise would, it is suggested, be to eliminate the element of *mens rea* as an element in criminal liability and to return to the old rule of strict or absolute liability.

In developing his argument Dr. Turner roundly asserts that negligence is a state of mind. It is 'the state of mind of a man who pursues a course of conduct *without adverting at all* to the consequences'.[29] Dr. Turner admits that this state of mind may be 'blameworthy'[30] and ground *civil* liability. Here it is important to pause and note that if anything is 'blameworthy', it is not the 'state of mind' but the agent's failure to inform himself of the facts and so *getting into* this 'state of mind'. But, says Dr. Turner, 'negligence, in its proper meaning of inadvertence cannot at Common Law amount to *mens rea*',[31] for 'no one could reasonably contend that a man, in a fit of inadvertence, could make himself guilty of the following crimes, "arson", "burglary", "larceny," "rape," "robbery" ...'[32] This of course is quite true; but proves nothing at all, until it is independently shown that to act negligently is the same as to act in 'a fit of inadvertence'. Precisely the same comment applies to the use made by Dr. Turner of many cases[33] where the judges have insisted that for criminal responsibility 'mere inadvertence' is not enough.

It is of course most important at this point to realize that the issue here is *not* merely a verbal one which the dictionary might settle. Much more is at stake; for Dr. Turner is really attempting by the use of his definitions to establish his general

[28] Ibid., p. 231. [29] Ibid., p. 207. [30] Ibid., p. 208. [31] Ibid., p. 209. [32] Ibid.

[33] E.g., *R.* v. *Finney* (1874), 12 Cox 625. See also *R.* v. *Bateman, Andrews* v. *D.P.P.,* and others discussed op. cit., pp. 216–17.

doctrine that if a man is to be held criminally responsible he must 'have in his mind the idea of bodily harm to someone', by suggesting that the only alternative to this is the quite repugnant doctrine that a man may be criminally liable for mere inadvertence when, through no failure of his to which the criminal law could attach importance, his mind is a mere blank. This alternative indeed would threaten to eliminate the doctrine of *mens rea*. But we must not be stampeded into the belief that we are faced with this dilemma. For there are not just two alternatives; we can perfectly well both deny that a man may be criminally responsible for 'mere inadvertence' and also deny that he is only responsible if 'he has an idea in his mind of harm to someone'. Thus, to take the familiar example, a workman who is mending a roof in a busy town starts to throw down into the street building materials without first bothering to take the elementary precaution of looking to see that no one is passing at the time. We are surely not forced to choose, as Dr. Turner's argument suggests, between two alternatives: (1) Did he have the idea of harm in his mind? (2) Did he merely act in a fit of inadvertence? Why should we not say that he has been grossly negligent because he has failed, though not deliberately, to take the most elementary of the precautions that the law requires him to take in order to avoid harm to others?

At this point, a careful consideration is needed of the differences between the meaning of expressions like 'inadvertently' and 'while his mind was a blank' on the one hand, and 'negligently' on the other. In ordinary English, and also in lawyers' English, when harm has resulted from someone's negligence, if we say of that person that he has acted negligently we are not thereby *merely* describing the frame of mind in which he acted. 'He negligently broke a saucer' is not the same *kind* of expression as 'He inadvertently broke a saucer'. The point of the adverb 'inadvertently' *is* merely to inform us of the agent's psychological state, whereas if we say 'He broke it negligently' we are not merely adding to this an element of blame or reproach, but something quite specific, viz. we are referring to the fact that the agent failed to comply with a standard of conduct with which any ordinary reasonable man *could* and *would* have complied: a standard requiring him to take precautions against harm. The word 'negligently', both in legal and in non-legal contexts, makes an essential reference to an omission to do what is thus required: it is not a flatly descriptive psychological expression like 'his mind was a blank'.

By contrast, if we say of an agent 'He acted inadvertently', this contains no implications that the agent fell below any standard of conduct. Indeed it is most often proffered as an excuse. 'X hit Smith inadvertently' means that X, in the course of doing some other action (e.g. sweeping the floor) through failing to attend to his bodily movements (e.g. his attention being distracted) and *a fortiori* not foreseeing the consequences, hit Smith.

There is of course a *connexion*, and an important one, between inadvertence and negligence, and it is this. Very often if we are to comply with a rule or standard requiring us to take precautions against harm we must, before we act, acquire certain information: we must examine or *advert* to the situation and its possible dangers (e.g. see if the gun we are playing with is loaded) and watch our bodily movements (handle the gun carefully if it is loaded). But this connexion far from identifying the concepts of negligence and inadvertence shows them to be different. *Through* our negligence in not examining the situation before acting or in attending to it as we act, we may fail to realise the possibly harmful consequences of what we are doing and as to these our mind is in a sense a 'blank'; but the negligence does not, of course, consist in this blank state of mind but in our failure to take precautions against harm by examining the situation. Crudely put, 'negligence' is not the name of 'a state of mind' while 'inadvertence' is.

We must now confront the claim made by Dr. Turner that there is an absurdity in stipulating that a special (gross) degree of negligence is required. 'There can be no different degrees of inadvertence as indicating a state of mind. The man's mind is a blank as to the consequences in question; his realization of their possibility is nothing and there are no different degrees of nothing'.[34] This *reductio ad absurdum* of the notion of gross negligence depends entirely on the view that negligence is merely a name for a state of mind consisting in the absence of foresight of consequences. Surely we should require something more to persuade us to drop notions so firmly embedded, not only in the law, but in common speech, as 'very negligent', 'gross carelessness', a 'minor form of negligence'. Negligence is gross if the precautions to be taken against harm are very simple, such as persons who are but poorly endowed with physical and mental capacities can easily take.[35] So, in the workman's case, it was gross negligence not to look and see before throwing off the slates; perhaps it was somewhat less gross (because it required more exertion and thought) to have failed to shout a warning for those not yet in view; it was less gross still to have failed to have put up some warning notice in the street below.

4. NEGLIGENCE AND NORMAL CAPACITIES

At the root of Dr. Turner's arguments there lie, I think, certain unexamined assumptions as to what the mind is and why its 'states' are relevant to responsibility. Dr. Turner obviously thinks that unless a man 'has in his mind the idea of harm to someone' it is not only bad law, but morally objectionable, as a recourse to strict or absolute liability, to punish him. But here we should ask why, in or out of law courts, we should attach this crucial importance to foresight of consequences, to the 'having of an idea in the mind of harm to someone'. On what theory of responsibility is it that the presence of this particular item of mental furniture is taken to be something which makes it perfectly satisfactory to hold that

[34] Op. cit., p. 211.

[35] 'It is such a degree of negligence as excludes the loosest degree of care' quoted by Hewart C. J. in *R.* v. *Bateman* (1925), 19 Cr. App. Rep. 8.

the agent is responsible for what he did? And why should we necessarily conclude that in its absence an agent cannot be decently held responsible? I suspect, in Dr. Turner's doctrine, a form of the ancient belief that possession of knowledge of consequences is a sufficient and necessary condition of the capacity for self control, so that if the agent knows the consequences of his action we are bound to say 'he could have helped it'; and, by parity of reasoning, if he does not know the consequences of his action, even though he failed to examine or think about the situation before acting, we are bound to say that he could not have helped it.

Neither of these views are acceptable. The first is not only incompatible with what large numbers of scientists and lawyers and plain men now believe about the capacity of human beings for self control. But it is also true that there is nothing to compel us to say 'He could not have helped it' in *all* cases where a man omits to think about or examine the situation in which he acts and harm results which he has not foreseen. Sometimes we do say this and should say it; this is so when we have evidence, from the personal history of the agent or other sources, that his memory or other faculties were defective, or that he could not distinguish a dangerous situation from a harmless one, or where we know that repeated instructions and punishment have been of no avail. From such evidence we may conclude that he was unable to attend to, or examine the situation, or to assess its risks; often we find this so in the case of a child or a lunatic. We should wish to distinguish from such cases the case of a signalman whose duty it is to signal a train, if the evidence clearly shows that he has the normal capacities of memory and observation and intelligence. He may say after the disaster, 'Yes, I went off to play a game of cards. I just didn't stop to think about the 10.15 when I was asked to play'. Why, in such a case, should we say 'He could not help it—because his mind was a blank as to the consequences'? The kind of evidence we have to go upon in distinguishing those omissions to attend to, or examine, or think about the situation, and to assess its risks before acting, which we treat as culpable, from those omissions (e.g. on the part of infants or mentally deficient persons) for which we do not hold the agent responsible, is not different from the evidence we have to use whenever we say of anybody who has failed to do something 'He could not have done it' or 'He could have done it'. The evidence in such cases relates to the general capacities of the agent; it is drawn, not only from the facts of the instant case, but from many sources, such as his previous behaviour, the known effect upon him of instruction or punishment, etc. Only a theory that mental operations like attending to, or thinking about, or examining a situation are somehow 'either there or not there', and so utterly outside our control, can lead to the theory that we are *never* responsible if, like the signalman who forgets to pull the signal, we fail to think or remember. And this theory of the uncontrollable character of mental operations would, of course, be fatal to responsibility for even the most cold-blooded, deliberate action, performed by an agent with the maximum 'foresight'. For just as the signalman, inspired by Dr. Turner's argument, might say 'My mind was a blank' or 'I just forgot' or 'I just didn't think, I could not help not thinking' so the cold-blooded murderer might say 'I just decided to kill; I couldn't help deciding'. In the latter case we do not normally allow this plea because we know from the general history of the agent, and others like him, that he could have acted differently. This general evidence is what is relevant to the question of responsibility, not the mere presence or absence of foresight. We should have doubts, which now find legal expression in the category of diminished responsibility, even in the case of deliberate murder, if it were shown that in spite of every warning and every danger and without a comprehensible motive the agent had deliberately and repeatedly planned and committed murder. After all, a hundred times a day persons are blamed outside the law courts for not being more careful, for being inattentive and not stopping to think; in particular cases, their history or mental or physical examination may show that they could not have done what they omitted to do. In such cases they are not responsible; but *if* anyone is *ever* responsible for *anything*, there is no general reason why men should not be responsible for such omissions to think, or to consider the situation and its dangers before acting.

5. SUBJECTIVE AND OBJECTIVE

Excessive distrust of negligence and excessive confidence in the respectability of 'foresight of harm' or 'having the thought of harm in the mind' as a ground of responsibility have their roots in a common misunderstanding. Both oversimplify the character of the subjective element required in those whom we punish, if it is to be morally tolerable, according to common notions of justice, to punish them. The reason why, according to modern ideas, strict liability is odious, and appears as a sacrifice of a valued principle which we should make, if at all, only for some overriding social good, is not merely because it amounts, as it does, to punishing those who did not at the time of acting 'have in their minds' the elements of foresight or desire for muscular movement. These psychological elements are not *in themselves* crucial though they are important as aspects of responsibility. What is crucial is that those whom we punish should have had, when they acted, the normal capacities, physical and mental, for doing what the law requires and abstaining from what it forbids, and a fair opportunity to exercise these capacities. Where these capacities and opportunities are absent, as they are in different ways in the varied cases of accident, mistake, paralysis, reflex action, coercion, insanity, etc., the moral protest is that it is morally wrong to punish because 'he could not have helped it' or 'he could not have done otherwise' or 'he had no real choice'. But, as we have seen, there is no reason (unless we are to reject the whole business of responsibility and punishment) *always* to make this protest when someone who 'just didn't think' is punished for carelessness. For in some cases at least we may say 'he could have thought about what he was doing' with just as much rational confidence as one can say of any intentional wrongdoing 'he could have done otherwise'.

Of course, the law compromises with competing values over this matter of the subjective element in responsibility as it does over other matters. All legal systems temper their respect for the principle that persons should not be punished if they could not have done otherwise, i.e. had neither the capacity nor a fair opportunity to act otherwise. Sometimes this is done in deference to genuine practical difficulties of proof; sometimes it represents an obstinate refusal to recognize that human beings may not be able to control their conduct though they know what they are doing. Difficulties of proof may lead one system to limit consideration of the subjective element to the question whether a person acted intentionally and had volitional control of his muscular movements; other systems may let the inquiry go further and, in relation to some offences, consider whether the accused had, owing to some external cause, lost the power of such control, or whether his capacity to control was 'diminished' by mental abnormality or disease. In these last cases, exemplified in 'provocation' and 'diminished responsibility', if we punish at all we punish *less*, on the footing that, though the accused's capacity for self-control was not absent its exercise was a matter of abnormal difficulty. He is punished in effect for a failure to exercise control; and this is also involved when punishment for negligence is morally justifiable.

The most important compromise which legal systems make over the subjective element consists in its adoption of what has been unhappily termed the 'objective standard'. This may lead to an individual being treated for the purposes of conviction and punishment as if he possessed capacities for control of his conduct which he did not possess, but which an ordinary or reasonable man possesses and would have exercised. The expression 'objective' and its partner 'subjective' are unhappy because, as far as negligence is concerned, they obscure the real issue. We may be tempted to say with Dr. Turner that just because the negligent man does not have 'the thought of harm in his mind', to hold him responsible for negligence is *necessarily* to adopt an objective standard and to abandon the 'subjective' element in responsibility. It then becomes vital to distinguish this (mistaken) thesis from the position brought about by the use of objective standards in the application of laws which make negligence criminally punishable. For, when negligence is made criminally punishable, this itself leaves open the question: whether, before we punish, both or only the first of the following two questions must be answered affirmatively:

(i) Did the accused fail to take those precautions which any reasonable man with normal capacities would in the circumstances have taken?

(ii) Could the accused, given his mental and physical capacities, have taken those precautions?

One use of the dangerous expressions 'objective' and 'subjective' is to make the distinction between these two questions; given the ambiguities of those expressions, this distinction would have been more happily expressed by the expressions 'invariant' standard of care, and 'individualised conditions of liability'. It may well be that, even if the 'standard of care' is pitched very low so that individuals are held liable only if they fail to take very elementary precautions against harm, there will still be some unfortunate individuals who, through lack of intelligence, powers of concentration or memory, or through clumsiness, could not attain even this low standard. If our conditions of liability are invariant and not flexible, i.e. if they are not adjusted to the capacities of the accused, then some individuals will be held liable for negligence though they could not have helped their failure to comply with the standard. In *such* cases, indeed, criminal responsibility will be made independent of any 'subjective element', since the accused could not have conformed to the required standard. But this result is nothing to do with negligence being taken as a basis for criminal liability; precisely the same result will be reached if, in considering whether a person acted intentionally, we were to attribute to him foresight of consequences which a reasonable man would have foreseen but which he did not. 'Absolute liability' results, not from the admission of the principle that one who has been grossly negligent is criminally responsible for the consequent harm even if 'he had no idea in his mind of harm to anyone', but from the refusal in the application of this principle to consider the capacities of an individual who has fallen below the standard of care.

It is of course quite arguable that no legal system could afford to individualize the conditions of liability so far as to discover and excuse all those who could not attain the average or reasonable man's standard. It may, in practice, be impossible to do more than excuse those who suffer from gross forms of incapacity, viz. infants, or the insane, or those afflicted with recognizably inadequate powers of control over their movements, or who are clearly unable to detect, or extricate themselves, from situations in which their disability may work harm. Some confusion is, however, engendered by certain inappropriate ways of describing these excusable cases, which we are tempted to use in a system which, like our own, defines negligence in terms of what the reasonable man would do. We may find ourselves asking whether the infant, the insane, or those suffering from paralysis did all that a reasonable man would *in the circumstances* do, taking 'circumstances' (most queerly) to include personal qualities like being an infant, insane or paralysed. This paradoxical approach leads to many difficulties. To avoid them we need to hold apart the primary question (1) What *would* the reasonable man with ordinary capacities have done in these circumstances? from the second question (2), *Could* the accused with *his* capacities have done that? Reference to such factors as lunacy or disease should be made in answering only the second of these questions. This simple, and surely realistic, approach avoids difficulties which the notion of individualizing the standard of care has presented for certain writers; for these difficulties are usually created by the mistaken assumption that the only way of allowing for individual incapacities is to treat them as part of the 'circumstances' in which the reasonable man is supposed to be acting. Thus Dr. Glanville Williams said that if 'regard must be had to the make-up and circumstances of the particular offender, one would seem on a determinist view of conduct to be pushed to the conclusion that there is no standard of conduct at all. For if every characteristic of the individual is taken into account, including his heredity the conclusion is that he could not help doing as he did.'[36]

But 'determinism' presents no special difficulty here. The question is whether the individual had the capacity (inherited or not) to act otherwise than he did, and 'determinism' has no relevance to the case of one who is accused of negligence which it does not have to one accused of intentionally killing. Dr. Williams supports his arguments by discussion of the case of a motorist whom a blow or illness has rendered incapable of driving properly. His conclusion, tentatively expressed, is that if the blow or illness occurred long ago or in infancy he should not be excused, but if it occurred shortly before the driving in respect of which he is charged he should. Only thus, it seems to him, can any standard of conduct be preserved.[37] But there seems no need to make this extraordinary distinction. Again, the first question which we should ask is: What *would* a reasonable driver with normal capacities have done? The second question is whether or not the accused driver had at the time he drove the normal capacity of control (either in the actual conduct of the vehicle in driving or in the decision to engage in driving). If he was incapable, the date recent or otherwise of the causal origin of the incapacity is surely beside the point, except that if it was of long standing, this would suggest that he knew of it and was negligent in driving with that knowledge.

Equally obscure to me are the reasons given by Dr. Williams for doubting the efficacy of punishment for negligence. He asks, 'Even if a person admits that he occasionally makes a negligent mistake, how, in the nature of things, can punishment for inadvertence serve to deter?[38] But if this question is meant as an argument, it rests on the old, mistaken identification of the 'subjective element' involved in negligence with 'a

[36] *Criminal Law: The General Part* (1st edn.), p. 82. In the second edition (p. 101) this passage is replaced by the following: 'But if the notional person by whom the defendant is judged is invested with every characteristic of the defendant, the standard disappears. For, in that case, the notional person would have acted in the same way as the defendant acted".

[37] op.cit. (1st edn.), p. 84. This passage is omitted from the second edition.

[38] op. cit. (2nd edn.), p. 123.

blank mind', whereas it is in fact a failure to exercise the capacity to advert to, and to think about and control, conduct and its risks. Surely we have plenty of empirical evidence to show that, as Professor Wechsler has said, 'punishment supplies men with an additional motive to take care before acting, to use their faculties, and to draw upon their experience.'[39] Again there is no difficulty here peculiar to negligence, though of course we can doubt the efficacy of any punishment to deter any kind of offence.

I should add (out of abundant caution) that I have not been concerned here to advocate punishing negligence, though perhaps better acquaintance with motoring offences would convert me into a passionate advocate. My concern has been to show only that the belief that criminal responsibility for negligence is a form of strict or absolute liability, rests on a confused conception of the 'subjective element' and its relation to responsibility.

[39] loc. cit. *supra* p. 138. 6.

VII

PUNISHMENT AND THE ELIMINATION OF RESPONSIBILITY

I

IN Dostoevsky's novel *Crime and Punishment*, a minor character, Lebezyatnikov, who is described as 'a follower of the latest ideas', explains that 'in this age the sentiment of compassion is actually prohibited by science, and that is how they order things in England where they have political economy'. This is a mocking reference to the social philosophy compounded out of Utilitarianism and scientific rationalism which was then regarded, with some reason, as typically English. Among a nation of shopkeepers, this philosophy might rank as progressive and enlightened, but it was viewed by Dostoevsky as a contagion from the West, to be hated and feared. Its spread would, he thought, blind men to the realities of human nature and experience, and would pervert institutions which gave expression, however clumsily, to spiritual values of profound importance. Utilitarianism and science between them would transform the government of responsible human beings into the manipulation of things.

Among the institutions which Dostoevsky most wished to preserve from perversion by the social philosophy of the West was the institution of punishment; and his novels make real to us, in a way that no abstract statement ever can, a conception of punishment which, since he wrote, has come to occupy a much diminished place in the penal policy and practice of this country, as it has in every country where the reduction of crime is felt to be one of the most urgent social problems. This older conception of punishment is sharply distinguished from mere social hygiene: it does not make primary, as modern thought does, the reduction of crime or the protection of society from the criminal; instead it makes primary the meting out to a responsible wrongdoer of his just deserts. Dostoevsky passionately believed that society was morally justified in punishing people simply because they had done wrong; he also believed that psychologically the criminal needed his punishment to heal the laceration of the bonds that joined him to his society. So, in the end, Raskolnikov the murderer thirsts for his punishment. Many of us here today—perhaps most of us—may hate these ideas as useless obstructions to rational thought about the worst of our social problems. Perhaps some of us would wish to hasten the disappearance of these ideas not only from public policy and the criminal law, but from human consciousness altogether. Nonetheless, we still need to understand the moral and psychological appeal which these ideas have, for they have not disappeared yet nor have they been relegated wholly to the sphere of private moral censure. In an attenuated form they still have a place among the now complicated and partly inconsistent set of ideas that jostle together in the mind of an English judge when he sentences the criminals convicted in his court.

In this lecture I shall attempt to describe the position which these ideas have come to hold in our penal practices and policies. I shall take from Hobhouse, in whose memory these lectures are founded, a lucid statement of what he himself termed 'the new order of ideas' of punishment.

'Punishment', he wrote, 'is compelled to justify itself by its actual effect, on society, in maintaining order without legalizing brutality, on the criminal, in deterring him or in aiding his reform.'[1] And punishment 'is not, like reward, a part of ideal justice, it is a mechanical and dangerous means of protection which it requires the greatest wisdom and humanity to convert into an agency of reform.'[2]

For such an outlook, the moral justification for punishment lies in its effects—in its contribution to the prevention of crime and the social readjustment of the criminal. It is essentially forward-looking: it considers the future good we can do to society including the criminal. In the pursuit of these

[1] *Morals in Evolution* (3rd edn.), p. 130.
[2] *The Elements of Social Justice*, p. 128.

forward-looking aims, we need all the resources of reason, experience and science; we cannot here be guided by our intuitions; we cannot know by looking only at what the criminal has done, what should be done to him. His act, lying in the past, is important merely as a symptom—one symptom among others—of his character, mind and disposition; it helps us to diagnose what he is like and predict the effects of our action on him and on society. So, if what we do to him is to advance any rational social purpose, we must understand his crime; but it cannot by itself dictate the kind or severity of punishment.

For this forward-looking Utilitarian approach the traditional conception of punishment has always presented difficulties. At two points, traditional punishment looks backward not forward. One of these points corresponds to the conviction by a Court, and the other to the Court's sentence. At both these stages the criminal's act is something more than a symptom on which diagnosis and prognosis may be based. It has an altogether different status. At the conviction stage, if punishment is to be justified at all, the criminal's act must be that of a responsible agent: that is, it must be the act of one who *could have* kept the law which he has broken. And at the sentence stage, the punishment must bear some sort of relationship to the act: it must in some sense 'fit' it or be 'proportionate' to it. Only if both these backward-looking requirements are satisfied can it be said that this man deserved punishment for this offence. These two backward-looking requirements are in fact closely related and because scepticism about one leads to scepticism about the other, I have chosen to discuss them both under the title of 'The Elimination of Responsibility'. But in fact it has only comparatively recently been openly urged that we should, while retaining a system of criminal law, eliminate legal responsibility or allow it to wither away; whereas the impact of penal reform on the old idea that the punishment should be fitted to the crime, rather than the criminal, has been much longer sustained and more severe. So although conviction comes before sentence I shall discuss the sentence aspect first.

2

Like many other features of punishment the idea of a sentence 'fitted' to a crime is susceptible of many different interpretations; as Bentham said, words like 'fit' or 'proportionate' are 'more oracular than instructive'. In its crudest form it is the notion that what the criminal has done should be done to him, and wherever thinking about punishment is primitive, as it often is, this crude idea reasserts itself: the killer should be killed; the violent assailant should be flogged. But as Blackstone pointed out—he was a reformer surprisingly enough on this aspect of punishment—this idea is incapable of generalization without absurdity. 'There are very many crimes, that will in no shape admit of these penalties, without manifest absurdity and wickedness. Theft cannot be punished by theft, defamation by defamation, forgery by forgery, adultery by adultery . . .'[3] Moreover, as he adds, even in the crude cases where the penalty seems to be dictated by 'natural reason' we may be deluded: there may appear to be an exact 'correspondence' in the case of the death penalty for murder, but the punishment may yet not be really equivalent to the crime. As he somewhat quaintly says, 'the execution of a needy decrepit assassin is a poor satisfaction for the murder of a nobleman in the bloom of his youth, and full enjoyment of his friends, his honours, and his fortune'.[4] Yet once we leave this crude version of the principle, things become difficult. The first refinement is the idea that the suffering imposed by punishment should be in some sense equal to or proportionate to the wickedness of the crime. But in what sense? How measure either wickedness or suffering in the absence of units of either? Even if we had more than the limited insight which is available to human judges into a criminal's motives, powers and temptations, there is no natural relationship to be discerned between wickedness and punishment of a certain degree

[3] Blackstone, *Commentaries,* Book IV, Chap. I; II.3 [4] Ibid.

or kind, so that we can say the latter naturally 'fits' the former. Those who see these difficulties and yet insist that punishment must somehow be related to wickedness or 'culpability' present their principle in a different form: what is required is not some ideally appropriate relationship between a single crime and its punishment, but that on a scale or tariff of punishments and offences, punishments for different crimes should be 'proportionate' to the relative wickedness or seriousness of the crime. For though we cannot say *how* wicked any given crime is, perhaps we can say that one is *more* wicked than another and we should express this ordinal relation in a corresponding scale of penalties. Trivial offences causing little harm must not be punished as severely as offences causing great harm; causing harm intentionally must be punished more severely than causing the same harm unintentionally.

Of course even this way of relating the severity of punishment to the seriousness of the crime is beset by many difficulties if we attempt to apply it at all literally. The first of these is relatively trivial: even if it were possible to arrange all crimes on a scale of relative seriousness, our starting point or base of comparison must be a crime for which a penalty is fixed otherwise than by comparison with others. We must start somewhere, and in practice the starting point is apt to be just the traditional or usual penalty for a given offence.[5] Secondly it is not clear what, as between the objective harm caused by a crime and the subjective evil intention inspiring it, is to be the measure of 'seriousness'. Is negligently causing the destruction of a city worse than the intentional wounding of a single policeman? Or are we to pay attention to both objective harm done and subjective wickedness? Thirdly if the subjective wickedness of the criminal act is relevant, can human judges discover and make comparisons between the motives, temptations, opportunities and wickedness of different individuals? No doubt if we consider *types* of crime—average cases—only loosely representative of actual individual crimes, a rough

[5] See Grünhut, *Penal Reform*, p. 5, and *Report of Departmental Committee on Persistent Offenders* (1932), p. 10.

approximation can be made to the idea of adapting the severity of punishment to the different 'culpability' or seriousness of different offences. We can lay down a few rough discriminations between intentional and unintentional injury: we can recognize standard types of temptation and weakness, and use these to mitigate or aggravate the severity of the standard punishment for a given type. We shall consider later the social purpose of such a rough conventional tariff. But it is well to remember its roughness. As Sir Ernest Gowers said, 'The fact is that the considerations that determine the culpability of any crime are infinite in number and variety, depending on the criminal as well as the crime, and cannot possibly be catalogued objectively without gross error in the application of the definition to particular cases.'[6]

So undoubtedly there is, for modern minds, something obscure and difficult in the idea that we should think in choosing punishment of some right intrinsic relation which it must bear to the wickedness of the criminal's act, rather than the effect of the punishment on society and on him.

Bentham himself offered a quite different theory of relative proportion between penalties and crime within a Utilitarian framework of individual and general prevention. Here the guiding considerations were forward-looking principles: minor harms were not to be punished with a severity that created greater misery than the offence unchecked; the potential offender must be supplied with an inducement in the shape of lesser penalties to commit a less harmful offence rather than a more harmful one. We may well think that Bentham's complex doctrines rested on an uncritical belief in the part played by calculation in anti-social behaviour. Nonetheless we may think the general direction right; whereas the obscure notions of fitness, equivalence or proportion to the act done seem to turn our thoughts uselessly in the wrong direction. Even Plato thought that looking back to the past deed (except as a symptom indicating what was likely to cure or prevent) was irra-

[6] *A Life for a Life*, p. 38.

tional. To measure punishment by reference to it was, he said, like 'lashing a rock'.[7]

But what part has this backward-looking idea, measuring punishment by wickedness, played in our policies and how has it been combined with the Utilitarian forward-looking elements in our penal practices?

In answering this question it is important to remember the variety and complexity of our penal institutions. We have first the action of the legislature fixing only maximum penalties[8] for crime, except in very rare cases, such as murder when there is no alternative penalty. Then we have the announced plans and policies of the executive responsible for prison administration. Thirdly, we have the actual practices of judges choosing sentences in the exercise of the vast discretion which for more than 100 years has been left to them. Here, it is safe to say, the ideas of fitness and proportion have had their fullest play, though it is true that the history of this fascinating side of the law has still to be written and it is extraordinarily difficult to find out how judges have viewed their task. What is clear is this: until the turn of the century there was nothing to place on our judges the duty of choosing penalties designed for reform or rehabilitation, or of adapting the punishment in any other way to the needs or potentialities of the criminal. There was nothing to stop a judge giving exclusive attention, if he wished to, to the wickedness of the crime and allowing his estimate of it to determine his sentence. Of course there were great penal reforms in the nineteenth century, but they were largely reforms not of what judges were to do in sentencing but what prisons were to do to those whom the judges had sentenced. The great maxims of the reformers—individualization of punishment, rehabilitation, training—so far as they made headway in the last century did so primarily inside the prisons in altering the life lived there by prisoners. The judges took no part in this and their sentencing powers were for long unaffected by it. Their operations were thought of as

[7] *Protagoras*, 324.

[8] See pp. 115–16 *supra*.

something given; and the new penal ideas worked within the given framework.[9]

Of course we know a good deal about the opposition which the reformers met in the last century. But it seems clear that the main opposition to ideas of re-education and reform came less from those who stressed retributive ideas of proportion to wickedness than from another wing in the Utilitarian camp: those who stressed the need to make punishment a general deterrent by example. The argument that milder reforming treatment might reduce recidivism was often met with the Utilitarian reminder: that to focus on the actual offender might jeopardize what was believed to be the quantitatively more important influence which the threat of severe punishment might have on potential offenders. And of course the dreary Utilitarian principle of 'lesser eligibility' figured frequently in the arguments over prison policy. Conditions inside must not be made preferable to life in the stinking alleys from which the prisoner came.

As long as the sentencing powers of judges were left outside the scope of reform there was little a judge could do, even if he were so minded, to adapt his sentence to reforming aims. But the series of great measures which in the end followed the Gladstone Report of 1895 brought about an immense change in the judges' powers and responsibilities. For the first time they were charged with the duty of considering the suitability of a sentence whose aim was sharply differentiated from retribution for past wickedness. The possibilities now included (besides what we may call ordinary imprisonment), probation, borstal training, corrective training and preventive detention, and wide powers of absolute and conditional discharge. So individualized treatment within prisons, adapted at least in theory to the needs and potentialities of the prisoner, was now complemented by the idea of individualized sentence; and the judges were made to participate in an activity which in the main had been a matter for administrators.

[9] See G. Rose, *The Struggle for Penal Reform*, esp. Chap. XVIII.

What signs of a clash were there and are there between these forward-looking ideas and the old ideas of proportion? Certainly there was some judicial restiveness, but in considering it, it is well to remember that the new individualized measures (particularly preventive detention for habitual criminals) did not always mean a sentence which criminals themselves liked better: very often it meant for them something much longer than what they would get from judges operating rough ideas of proportion. The strength of the old idea that the amount of punishment was, in spite of the new individualizing aims, something which ought to be measured by the crime showed itself in the way in which the earliest form of preventive detention was introduced in 1908. The system known on the Continent as the 'double track' was adopted. The habitual offender who satisfied the requirements for this form of treatment was to receive two sentences: one was the punishment fixed on conventional lines for his last offence; the other was the special individuated measure. So his crime had two aspects: one was that of a responsible piece of wickedness deserving as such a certain punishment; the other was that of a symptom of disposition or character and an index of appropriate preventive treatment. On the Continent the 'double-track' system has been elaborated in ways which may seem to us somewhat metaphysical: punishment which is to be 'guilt adequate', i.e. orientated towards the criminal act, is carefully distinguished from mere 'measures' orientated to the criminal's character and the needs of society.[10] The recent German Penal Code preserves this distinction though it is regretted as artificial by many.[11] Certainly the prisoner who after serving a three-year sentence is told that his punishment is over but that a seven-year period of preventive detention awaits him and that this is a 'measure' of social protection, not a punishment, might think he was being tormented by a barren piece of

[10] See '"Rule of Law" in Criminal Justice', by H. Silving, in *Essays in Criminal Science* (1961), ed. J. Mueller, pp. 98 et seq.

[11] See 'German Criminal Law Reform', by H-H. Jescheck, in op. cit., ed. Mueller, p. 395.

conceptualism—though he might not express himself in that way.

But these expedients, or—as some may think of them—subterfuges, are signs of the vitality of the old ideas. If we look for signs of their persistence in current English judicial practice we can find them easily enough. Though the double-track system has been abandoned here, our judges have always felt uneasy when faced with a conflict between what they consider to be a punishment appropriate to the seriousness of a crime, and the steps which one of the individualized forms of punishment might require. Sometimes this emerges into the light of day in reported cases. Thus it is now the law that a young offender may be sent to borstal training which may last as long as three years, although his last offence is punishable by a maximum penalty of one year. But, for many years, courts of first instance have refused to do this and the Court of Criminal Appeal upheld them in this until last year when, by a sudden reversal of principle, hard indeed to reconcile with a doctrine of binding precedent, the offender's last offence was allowed to figure as a symptom of the need for reformative treatment rather than as determining by itself the measure of punishment.[12]

But we now have other evidence of the importance attached by English judges to the idea that punishment should be proportioned to the seriousness of the offence. The recent report of the Interdepartmental Committee on the Business of the Criminal Courts (known after its chairman as the Streatfield report)[13] makes quite plain how prominent a place in sentencing the idea of a tariff has had and indeed still has. It gives us first a glimpse into the recent past: 'Sentencing used to be a comparatively simple matter. The primary objective was to fix a sentence proportionate to the offender's culpability, and the system has been loosely described as the "tariff system". . . . In addition, the courts have

[12] *R.* v. *Amos* (1960), 45 Cr. App. Rep. 42. See also *R.* v. *McCarthy* (1955), 39 Cr. App. Rep. 118, and *R.* v. *Ball* (1951), 35 Cr. App. Rep. 164.

[13] *Report of the Interdepartmental Committee on the Business of the Criminal Courts*, Cmd. 1289 (1961).

always had in mind the need to protect society from the persistent offender, to deter potential offenders and to deter or reform the individual offender. But in general it was thought that the "tariff system" took the three other objectives in its stride: giving an offender the punishment he deserved was thought to be the best way of deterring him and others and of protecting society.'[14] That this tariff idea has overshadowed forward-looking measures seems clear from the Committee's statement that the key to advance in this field is to recognize the fundamental difference between assessing culpability and the other objectives of sentencing. The Committee itself indeed emphasized that the offender's culpability is still a factor to be weighed in the choice of sentence; but its main recommendations are designed to secure that Courts should be equipped with information which will enable them to make an intelligent choice between the forward-looking measures now available. Many of the judges who gave evidence expressed satisfaction with things as they are: but the Committee expressed the view that there are now 'difficulties and anomalies' in the supply of information to the Courts and that 'sentencers are provided with inadequate information about what the various forms of sentence involve and achieve'.[15]

It is indeed difficult not to conclude that the satisfaction with things as they are, expressed by so many witnesses must be due in large part to the fact that the idea of a 'tariff' is still for many English judges a primary determinant of punishment. And of course psychologically there is every reason why they should be inclined to this view. When an English judge reaches the bench and assumes the responsibilities of sentencing he may have spent many arduous years dealing with civil or commercial cases. Very few new judges will have had recent contact with criminal law or criminals, or have practised regularly in the criminal courts, though many will have sat as Recorders for perhaps ten days a year in the intervals of a busy practice. There is little or nothing in the usual social background of an English barrister or in his education or training or (unless he specializes in criminal work) in his

[14] Ibid., paras. 257–8. [15] Ibid., para. 265.

career, to give him any particular understanding of the social conditions, and the psychology of criminals, of the methods or aims of new penal measures or of the relevant social sciences. What he will have are qualities which make an English judge an incomparable master of the art of a fair trial: an exactly trained sense of justice and fairness and great common sense. It is altogether natural that he should give primacy to that aspect of sentencing which relies most on these qualities and least on the technical or expert knowledge which he has not got. But though natural it does not follow that this is as it should be.

But there is something more than these somewhat inarticulate practices to go on. We have in England an articulate judicial doctrine or reinterpretation of the idea that a punishment must fit a crime. Nowhere has this been more powerfully expounded than by the great Victorian judge James Fitzjames Stephen in his history of the criminal law and his book *Liberty, Equality, Fraternity* written in reply to Mill on Liberty. Of course not all judges share his views, but in our own day very much the same views have been expressed by distinguished judges. Stephen, like his successors today, emphasized the interdependence and interpenetration of law and morals, not only as a fact to be observed but as something we should foster and intensify. In his view the object of the criminal law could not be stated either in the old Utilitarian language of deterrence or in the newer language of reform and rehabilitation. He insisted that the criminal law did and should operate to 'give distinct shape' to moral indignation, and hatred of the criminal. Because the criminal law has this function 'Everything which is regarded as enhancing the moral guilt of a particular offence is recognized as a reason for increasing the severity of the punishment . . . the sentence of the law is to the moral sentiment of the public what a seal is to hot wax'.[16] This is of course strong stuff and it would be surprising to find our judges repeating today Stephen's assertions concerning the moral rectitude of *hating* criminals, nor would they speak as Stephen did about the

[16] *A History of the Criminal Law in England,* Vol. II, Chap. XVII, p. 81.

duty of the judge to express that desire for vengeance which crime excites in a healthy mind.[17] Nonetheless Stephen's view of the relation of criminal law to morality and what may be called his expressive or denunciatory theory of criminal punishment is with us still. Thus Lord Denning told the Royal Commission on Capital Punishment that 'the punishment for grave crimes should adequately reflect the revulsion felt by the great majority of citizens for them. It is a mistake', he said, 'to consider the object of punishment as being deterrent or reformative or preventive and nothing else. . . . The ultimate justification of any punishment is not that it is a deterrent, but that it is the emphatic denunciation by the community of a crime: and from this point of view there are some murders which, in the present state of public opinion, demand the most emphatic denunciation of all, namely the death penalty'.[18]

What are the merits of this denunciatory theory of punishment? Does it offer an acceptable interpretation of what makes a punishment fit a crime? Or does it exaggerate the claim that that idea should have on the attention of judges passing sentence? Again, because of their training and situation, this view of punishment is one which usually is likely to commend itself to most English judges; for, in exercising the great discretion which the law leaves to them in sentencing, they wish neither to appear to be acting arbitrarily nor to be applying abstract or scientific penological techniques of which they have little knowledge or experience. It is therefore natural for them to claim to be the mouthpiece of the moral sentiments of society. Yet I think we should distrust this theory for at least three reasons:

First. One has every sympathy with judges in their difficult task but this theory makes things altogether too easy both for judges and the community. To tell judges that the expression

[17] But Lord Goddard said in the Lords debate on capital punishment in 1956: 'I do not see how it can be said to be either non-Christian or can be regarded in any other way than as praiseworthy that the country should be willing to avenge crime.' 198 *H.L. Deb.*, 743 (1956).

[18] *The Royal Commission on Capital Punishment*, Cmd. 8932, para. 53.

of the community's moral indignation is 'the ultimate justification of punishment' is to tempt them from the task of acquiring knowledge of and thinking about the effects of what they are doing. And for the community to think that there is something sacrosanct about its scale of moral evaluations may, as Mill long ago told us, stultify its advance. For these evaluations may plainly rest on inadequate understanding or appreciation of facts. Should we, for example, wish common estimates of the moral seriousness of bad driving on the roads to be taken as the prime determinant of punishments? Or should we hope that the law might, here and elsewhere, not passively reflect uninstructed opinion but actively help to shape moral sentiments to rational common ends?

Secondly. The operation of this theory may be very deceptive. It is true that for all social moralities certain major evaluations hold good. Intentional killing is worse than unintentional killing; inflicting minor harm is less bad than inflicting major harm; wounding is worse than stealing apples. But it is sociologically very naïve to think that there is even in England a single homogeneous social morality whose mouthpiece the judge can be in fixing sentence, and in admitting one thing and rejecting another as a mitigating or aggravating factor. Our society, whether we like it or not, is morally a plural society; and judgments of the relative seriousness of different crimes vary within it far more than this simple theory recognizes.[19] Judges talk much of the judgments of the 'ordinary reasonable man' and claim to be able to discover what he thinks. But the method used is usually introspection and this is because the judgment of the reasonable man very often is a mere projected shadow, cast by the judge's own moral views or those of his own social class.

Thirdly. It is indeed important that the law should not in its scale of punishments gratuitously flout any well-marked

[19] Research into these matters is still in its infancy but there is little to suggest that there is much general agreement. See the results of the B.B.C. Audience Research quoted by J. Silvey in 'The Criminal Law and Public Opinion' in *Crim. L.R.* (1961), at p. 355. No offence was agreed to be the 'most serious' by more than one in four.

common moral distinctions. But this is because the claim of justice that 'like cases should be treated alike' should always be heard; and where the law appears to depart from common estimates of relative wickedness it should make clear what moral aims require this. The double track system at least had the merit of doing this. But respect for justice between different offenders expressed in the injunction 'treat like cases alike' is not to be identified with a denunciatory theory of punishment; it does not involve search for a penalty the severity of which, like a gesture, will aptly express a specific feeling of revulsion or moral indignation. This is indeed a semi-aesthetic idea which has wandered into the theory of punishment. Surely to think of the apt expression of feeling—even if we call it moral indignation rather than revenge—as the ultimate *justification* of punishment is to subordinate what is primary to what is ancillary. We do not live in society in order to condemn, though we may condemn in order to live. On the other hand the injunction 'treat like cases alike' with its corollary 'treat different cases differently' has indeed a place as a prima facie principle of fairness between offenders, but not as something which warrants going beyond the requirements of the forward-looking aims of deterrence, prevention and reform to find some apt expression of moral feeling. Fairness between different offenders expressed in terms of different punishments is not an end in itself, but a method of pursuing other aims which has indeed a moral claim on our attention; and we should give effect to it where it does not impede the pursuit of the main aims of punishment. Very often English judges have looked upon claims to equal treatment for different offenders in just this light. When a crime has become exceptionally or dangerously frequent judges have defended punishing an offender more severely than previous offenders on the ground that this step is necessary to check a major evil. It is well that this and other sacrifices of principles of equality between different offenders should be made only with hesitation and with full explanation; for there is always great danger that they may be made in moments of panic or without reliable evidence that they will prevent a worse evil. The idea of proportion interpreted in this way—as respect for a principle of fairness between different offenders—has still a place in an account of the values which a theory of punishment should recognize. But it is a modest place, and judging by the available evidence it is different from that assigned to ideas of proportion in much current judicial practice and theory.[20]

3

So much then for the backward-looking aspect of the sentence, fitting the punishment to the past crime rather than to future needs either of society or of the criminal himself. If we are sceptical about claims that this is a primary object of punishment, how does the scepticism bear on the other backward-looking aspect of punishment—conviction by a court? Here the conventional doctrine says that if punishment is to be justified at all it must be punishment for a responsible act. This is a more complex topic and is made so by three things. First, the idea of personal responsibility—indeed the very word 'responsibility'—is no less susceptible of different interpretations than the idea of a punishment fitting a crime. Secondly, discussions of this topic are often clouded by philosophical assumptions, sometimes concealed, especially those which concern determinism and free will. Thirdly, the extent

[20] It is important to distinguish the principle of justice that offences of different gravity be treated differently, from the general principle that trivial offences should not be punished with great severity, which rests on the simple Utilitarian ground that the law should not inflict greater suffering than it is likely to prevent. Conflicts between both these principles and modern reformative or preventive methods (whether called 'treatment' or punishment) are perhaps most acute in the juvenile court. Where magistrates consider the home background to be bad, conviction for a petty offence may be the occasion for removing a child from his home for a long period. This may appear to both child and parents as a disproportionate sentence for a crime, even if intended as a measure of welfare or protection. Further, if there are grounds for doubting the efficacy of such methods ('they turn out the same in the end whatever you do') the moral case for administering what is in effect a severe penalty is correspondingly doubtful. See W. E. Cavanagh, *The Child and the Court*, esp. Chaps. VII and VIII, and the *Report of the Committee on Children and Young Persons* (Cmd. 1191) (the 'Ingleby Report'), esp. paras. 60–66.

and the forms in which our legal system has recognized this principle have varied at different periods in quite complicated ways.

What is clear is at least this. For some centuries English law, like most civilized legal systems, has made liability to punishment for serious crime depend, not only on the accused doing the outward acts which the law forbids, but on his having done them in certain conditions which may broadly be termed mental. These mental conditions of responsibility are commonly referred to by lawyers as *mens rea*. This has meant that, subject to certain important qualifications, liability to punishment is excluded if the law was broken unintentionally, under duress or by a person judged to be below the age of responsibility or to be suffering from certain types of mental disease. But nowhere on its face does English law explain why these conditions are required, and it does not indeed refer to any general requirement of a responsible or voluntary act; only recently has the word 'responsibility' crept into our criminal statutes. For the most part English law merely recognizes as excuses the absence of certain crucial mental elements. Nonetheless, most lawyers, laymen and moralists, considering the legal doctrine of *mens rea* and the excuses that the law admits, would conclude that what the law has done here is to reflect, albeit imperfectly, a fundamental principle of morality that a person is not to be blamed for what he has done if he could not help doing it. This is how Blackstone at the beginning of modern legal history looked at the various excuses which the law accepted. He said they were accepted because 'the concurrence of the will when it has its choice either to do or avoid the act in question, [is] the only thing that renders human actions praiseworthy or culpable'.[21]

But now a number of factors complicate the story and make debatable the moral basis of this requirement. The first is that though the law approximates in its doctrine of the mental conditions of responsibility to what the moralist requires for moral blame, it is an approximation only and not a complete

[21] *Commentaries,* Book IV, Chap. II.

convergence. One reason why this is so is that the law has to compromise with a number of different claims, besides the claim that liability to punishment must be dependent on a voluntary responsible act. Proof of mental elements—especially to juries—is a difficult matter, and the law has often abandoned the attempt to discover whether a person charged with crime actually intended to do it, and has used instead certain presumptions, such as that a man intends the natural consequences of his action, or has used what are called objective tests, so that it is enough for conviction that an ordinary man who behaved as the accused did, would have foreseen certain consequences. Perhaps I need not remind you that the law still takes this view so far as murder is concerned, and the objective test of intention has recently received the approbation of the House of Lords, so that in the Lord Chancellor's words, 'once the jury are satisfied that the accused was unlawfully and voluntarily doing something to someone it matters not what the accused in fact contemplated as the probable result or whether he even contemplated at all'.[22]

But the law's reception of the principle that a person should not be liable to punishment except for a voluntary act has been limited in other ways. In its admission of excuses it has concentrated almost exclusively on lack of knowledge rather than on the defects of volition or will. So it has given only a minor place to the pleas of provocation, duress, and necessity, and only after a long struggle has it admitted, in cases of homicide, that a man may still not be fully responsible even if he knew what he did and that it was illegal. One reason for this is that the question whether a man did a given act knowingly or not is a question which, even when it is difficult, is simple compared with the question whether a man could have abstained from doing what he in fact did. The question

[22] *D.P.P.* v. *Smith* (1961) A.C. 290. But this is no longer law. See the Criminal Justice Act, 1967, s. 8 and *R.* v. *Wallett* (1968), 2 Q.B. 367.

whether a person had sufficient capacity or will power to conform to the law's requirements opens wider general issues and what shall count as evidence for or against is far from clear. Another factor making for concentration on cognitive elements in responsibility is the survival of the belief, in spite of psychological doctrines to the contrary, that if a man knows what he is doing, it *must* be true that he has a capacity to adjust his behaviour to the requirements of the law.

Thirdly, and this is a development of the last hundred years, the general principle that for criminal liability there must be *mens rea* has been qualified by the admission into the body of the law of a number of offences where liability is said to be 'strict'. For the most part these offences are petty and punished by fines. They concern the maintainance of standards in the manufacture of goods, though recently they have been extended to embrace more serious offences such as dangerous driving where penalties may be considerable. Where liability is strict it is no defence that the accused did not intend to do the act which the law forbids and could not, by the exercise of reasonable care, have avoided doing it.[23]

Strict liability is held in some considerable odium by most academic writers and by many judges. But why? What is so precious in the normal requirements of *mens rea* and how does this normal requirement fit together with our general aims in punishing? It is at this point that scepticism about the old idea that the primary measure of punishment is the wickedness of the criminal act leads to further scepticism about the importance of a responsible act as a condition of punishment. So long as punishment is viewed as a return of pain or suffering for moral evil done, justified by the intrinsic fitness of sentence to crime, or so long as a denunciatory theory is accepted, in which the ultimate justification of punishment is held to be its function as an expression of a community's moral indignation, it is easy to give an explanation of the importance of the requirement of a voluntary act as a condition

[23] See Glanville Williams, *Criminal Law: The General Part* (2nd edn.), Chap. VI for an account and criticism of 'Strict Responsibility' in English law.

of punishment. For in most western morality 'ought' implies 'can' and a person who could not help doing what he did is not morally guilty. But the case is altered if we no longer justify punishment in these ways; if we think of it as justified by its social aims and effects in protecting society and reforming the criminal, and if principles of proportion have only the minor place that I assigned to them. Either we must think again and find a new reason for requiring a responsible act as a condition of criminal punishment, or we must admit that there is no reason for retaining this feature of our institutions. Why should we not eliminate it altogether?

4

One major reason, then, for querying the importance of the principle of responsibility is the belief that it only has a place if punishment is backward-looking and retributive in aim. This is not the only reason for scepticism and I shall mention some others later. But I shall first attempt to assess the balance of forces as between conservatives and reformers on this issue. They are not so neatly aligned or divided as they are on the previous question of sentencing policy, and recent legal history on this matter does not show a continuous development but a movement of conflicting tendencies. Generally speaking, those who have taken a forward Utilitarian view of punishment have till recently also taken very seriously the notion of responsibility. On the whole their efforts have been directed to making more complete the law's imperfect recognition of this principle. They have tried to secure that enquiries into the state of mind of the accused person should be genuine and not matters of presumption or objective tests. Similarly, there has been great and partly successful pressure exerted upon the law to abandon old ideas of constructive crime and, especially in the case of mental disease, to shed its narrow concentration on cognitive elements. The most recent chapter in this history was the introduction in the Homicide Act of

1957 of the notion of diminished responsibility to supplement the old M'Naghten tests. In many states of America the movement took the overt form of requiring the law to find, as a condition of liability of a person suffering from mental disease, that he could have conformed to the law's requirements. Our own expedient has been a compromise, and a queerly worded one at that, in which 'impaired mental responsibility' is accepted, not as an excuse, but as reducing what would otherwise be murder to manslaughter not in any case punishable with death. So too, those who have taken a forward-looking view of the aims of punishment protested, though less successfully, against strict liability as a form of injustice which, apparently because of difficulties of proof, would convict, as guilty of the same crime, those who intended along with those who did not intend to do what the law forbids. These tendencies are evident in the work of such Utilitarian jurists as Dr. Glanville Williams in this country and Professor Herbert Wechsler of Columbia University in the United States.

But other voices are certainly to be heard within the camp of those who are united in rejecting retributive views of punishment and would agree that the principle that a punishment is required to be measured by moral guilt has only a subordinate part to play. Certain Utilitarian-minded thinkers share a scepticism about the whole idea that the courts can usefully enquire into whether or not a person could have done what he did not do, and to them the enlightened policy seems to be the one in which we by-pass this question: we should neither assert nor deny that the accused could have done otherwise than he did. Instead we should look upon his act merely as a symptom of the need for either punishment or treatment.[24] Such a proposal to eliminate responsibility may be more or less extreme. Some no doubt, at any rate as a first instalment, would confine it to cases where there is a prima facie evidence of mental abnormality, generously interpreted to include the psychopath. In this less extreme version the doctrine of *mens*

[24] See Barbara Wootton, *Social Science and Social Pathology*, esp. Chap. VIII; and 'Diminished Responsibility: A Layman's View', 76 *L.Q.R.* (1960) p. 224.

rea would be left in the law and 'normal' persons at least would continue to be excused if they could show that they acted unintentionally or under duress, &c., as they are at present Presumably the justification of this halfway measure would be that in most cases of mere accidental or unintentional conduct, or crimes committed under duress, the accused is not likely to repeat his crime and punishment or treatment is not necessary. Bentham indeed attempted to reinterpret in this way the whole doctrine of *mens rea*; the various 'mental excuses' admitted by the law were for him not *indicia* of the fact that the accused could not help what he did, but mere evidences that his punishment could not be useful to society.[25] But it seems clear that he was wrong in thinking that punishment of those who acted without *mens rea* could not be socially 'useful'. Strict liability may be objectionable on many different grounds but the utilitarian argument that it prevents evasion of the law by those who would be prepared to fabricate pleas of mistake or accident has some plausibility.[26]

But as I have said, many factors besides the decay of retributive ideas have contributed to the modern scepticism of the idea that a necessary condition of criminal punishment should be an act which the criminal could have avoided doing. Among these other factors is a certain interpretation of determinism. It is not unnatural that those who in general take a scientific approach to social problems should be powerfully influenced by what may be termed the ideology of science, and think that loyalty to scientific principles requires that we should abandon the idea that a man could have done something which he in fact did not do. Secondly, scientific ideology prompts the suggestion that what is in a man's mind—his power or capacity to resist temptation—not being open to observation by others is not something which we can discover by rational methods.[27]

But even if these two objections are waived many may be impressed by the great difficulty in specifying clear criteria

[25] *Principles of Morals and Legislation,* Chap. XIII.

[26] See Prolegomenon to the Principles of Punishment, Chap. I, *supra.*

[27] See Glanville Williams, 'Diminished Responsibility', in *Medicine, Science and Law* (Oct. 1960), p. 41, and Barbara Wootton, op. cit., 76 L.Q.R. (1960) pp. 232, 235.

by which we are to judge whether a person who acted in a certain way had the power or capacity to act otherwise. As Lady Wootton has claimed in her study of the cases of diminished responsibility, it often seems that the criteria formulated are circular or irrelevant. Sometimes an attempt is made to infer a lack of capacity to obey the law from the mere fact of repeated disobedience; sometimes the lack of capacity is illicitly deduced from the existence of a variety of mental conditions, depression or anger, and no convincing evidence of connexion exists. It is impossible not to sympathize with her conclusion that for the most part in these cases of mental abnormality all that we can do is to use the evidence to diagnose those whose mental abnormalities are likely to result in harmful conduct, and to predict what treatment will best prevent their repeating it. The moral that she draws is that the sooner we get down to this piece of honest toil instead of claiming to discover the undiscoverable the better.[28]

I shall not deal here with these objections or the details of such arguments as Lady Wootton's, designed to show that responsibility, in the sense of capacity to conform to the law's requirement, is not something whose presence or absence we can discover by rational means. Full consideration of the arguments involves quite complex logical issues, and I shall only say here that I do not regard all the arguments as conclusive. Here however I wish to reconsider the assumption, which seems to me to be very widespread, that only within the framework of a theory which sees punishment in a retributive or denunciatory light does the doctrine of responsilibity make sense. There is, I believe, at this point something to defend, a moral position which ought not to be evacuated as if the decay of retributive ideas had made it untenable. There are values quite distinct from those of retributive punishment which the system of responsibility does maintain, and which remain of great importance even if our aims in punishing are the forward-looking aims of social protection. Perhaps there is

[28] *Social Science and Social Pathology*, Chap. VIII.

something stale and outmoded in the terms in which we tend to discuss the morality of punishment—as if we were forced to choose between retribution and an Erewhon where we never raise the question 'could he help it?' What is needed is a re-interpretation of the notions of desert and responsibility, and fresh accounts of the importance of the principle that a voluntary act should normally be required as a condition of liability to punishment. Such a reinterpretation would not stress, as our legal moralists do, the importance of judgments of degrees of wickedness about which there is far less agreement than they suppose. Instead it would stress the much more nearly universal ideas of fairness or justice and of the value of individual liberty.

Thus a primary vindication of the principle of responsibility could rest on the simple idea that unless a man has the capacity and a fair opportunity or chance to adjust his behaviour to the law its penalties ought not to be applied to him. Even if we punish men not as wicked but as nuisances, this is something we should still respect. Such a doctrine of fair opportunity would not only provide a rationale for most of the existing excuses which the law admits in its doctrine of *mens rea* but it could also function as a critical principle to demand more from the law than it gives. That is, in its light we might question English law's general adherence to the doctrine that ignorance of the law does not excuse and in its light we might press further objections to strict liability.

But more could be said by way of reinterpretation of the principle of responsibility. Its importance emerges afresh if for the moment we imagine that we had eliminated this principle and changed to a system in which all liability was strict. What should we lose? Among other things, we should lose the ability which the present system in some degree guarantees to us, to predict and plan the future course of our lives within the coercive framework of the law. For the system which makes liability to the law's sanctions dependent upon a voluntary act not only maximizes the power of the individual to determine by his choice his future fate; it also maximizes his power to identify in advance the space which will be left open to him free from the law's interference. Whereas a system from which responsibility was eliminated so that he was liable for what he did by mistake or accident would leave each individual not only less able to exclude the future interference by the law with his life, but also less able to foresee the times of the law's interference.

Thirdly, there is this. At present the law which makes liability to punishment depend on a voluntary act calls for the exercise of powers of self-control but not for complete success in conforming to law. It is illuminating to look at the various excuses which the law admits, like accident or mistake, as ways of rewarding self-restraint. In effect the law says that even if things go wrong, as they do when mistakes are made or accidents occur, a man whose choices are right and who has done his best to keep the law will not suffer. If we contrast this system with one in which men were conditioned to obey the law by psychological or other means, or one in which they were liable to punishment or 'treatment' whether they had voluntarily offended or not, it is plain that our system takes a risk which these alternative systems do not. Our system does not interfere till harm has been done and has been proved to have been done with the appropriate *mens rea*. But the risk that is here taken is not taken for nothing. It is the price we pay for general recognition that a man's fate should depend upon his choice and this is to foster the prime social virtue of self-restraint.

Underlying these separate points there is I think a more important general principle. Human society is a society of persons; and persons do not view themselves or each other merely as so many bodies moving in ways which are sometimes harmful and have to be prevented or altered. Instead persons interpret each other's movements as manifestations of intention and choices, and these subjective factors are often more important to their social relations than the movements by which they are manifested or their effects. If one person hits another, the person struck does not think of the other as *just* a cause of pain to him; for it is of crucial importance to him whether the blow was deliberate or involuntary. If the blow was light but deliberate, it has a significance for the person struck quite different from an accidental much heavier blow. No doubt the moral judgments to be passed are among the things affected by this crucial distinction; but this is perhaps the least important thing so affected. If you strike me, the judgment that the blow was deliberate will elicit fear, indignation, anger, resentment: these are not voluntary responses; but the same judgment will enter into deliberations about my future voluntary conduct towards you and will colour all my social relations with you. Shall I be your friend or enemy? Offer soothing words? Or return the blow? All this will be different if the blow is not voluntary. This is how human nature in human society actually is and as yet we have no power to alter it. The bearing of this fundamental fact on the law is this. If as our legal moralists maintain it is important for the law to reflect common judgments of morality, it is surely even more important that it should in general reflect in its judgments on human conduct distinctions which not only underly morality, but pervade the whole of our social life. This it would fail to do if it treated men merely as alterable, predictable, curable or manipulable things.

For these reasons then I think there will be a place for the principle of responsibility even when retributive and denunciatory ideas of punishment are dead. But it is important to be realistic: to be aware of the social costs of making the control of anti-social behaviour dependent on this principle and to recognize cases where the benefits secured by it are minimal. We must be prepared *both* to consider exceptions to the principle on their merits *and* to be careful that unnecessary invasions of it are not made even in the guise of 'treatment' instead of frankly penal methods. At present we have in strict liability clear exceptions to the principle, but no very persuasive evidence that the sacrifice of principle is warranted here by the amount of dishonest evasion of conviction which would ensue if liability were not strict. On the other hand we may find that certain cases fall between the two stools of criminal punishment and medical treatment. In February 1961 a man charged with murder of a woman pleaded that he had killed her in his sleep.[29] He was acquitted and discharged. Though there is no reason to doubt the honesty of this man it is clear that every defence of this sort both opens the door to possible frauds and may mean freedom for dangerous persons. Yet it may be thought that both risks are too slight to warrant any alteration of the law.

The most difficult problems are presented by young children and the mentally abnormal. The latest proposals[30] for the treatment of young offenders are that the age of responsibility be raised from 8 to 12, and that measures hitherto applied to children under 12 in the exercise of a criminal jurisdiction to offenders judged responsible shall henceforth be applied in the exercise of a civil jurisdiction as measures of 'care and discipline' without any enquiry into their responsibility. It may be thought that the benefits to very young children of a system of responsibility are too small to weigh conclusively against the need to save them from a criminal career, even if the attempt to save them involves measures which may be difficult (in spite of the formal differences) for them or their parents to distinguish from punishment and in some cases from a heavy punishment for a trivial offence.

No less important are the provisions in the Mental Health Act, 1959, in relation to psychopaths: these empower the courts to send persons who have done what the law forbids either to prison or to hospital although there may be no real evidence *either* that they were on account of their mental condition incapable of acting otherwise than they did *or*, on the other hand, that like ordinary criminals they were capable of conforming to law.[31] This, as Lady Wootton has persuasively argued, is in substance to by-pass the question of responsibility. The wisdom and justice of these and other departures from

[29] See the case of Staff-Sergeant Boshears, *The Times*, 18 Feb. 1961.

[30] In the *Report of the Committee on Children and Young Persons* (Cmd. 1191).

[31] See the Mental Act, 1959, s. 60. This section does not apply to offences for which a penalty is fixed by law, e.g. murder. For compulsory detention in hospital under this section, the medical evidence required is that the offender is suffering from mental illness, psychopathic disorder, subnormality or severe subnormality and the mental disorder is such as to warrant detention in hospital.

the principle of responsibility may be debated. But I shall not debate them at this hour. My concern has only been to show that the principle of responsibility, which may be sacrificed when the social cost of maintaining it is too high, has a value and importance quite independent of retributive or denunciatory theories of punishment which we may very well discard.

VIII

CHANGING CONCEPTIONS OF RESPONSIBILITY

I

THIS lecture is concerned wholly with criminal responsibility and I have chosen to lecture on this subject here because both English and Israeli law have inherited from the past virtually the same doctrine concerning the criminal responsibility of the mentally abnormal and both have found this inheritance embarrassing. I refer of course to the M'Naghten rules of 1843. In Israel the Supreme Court has found it possible to supplement these exceedingly narrow rules by use of the doctrine incorporated in s. 11 of the Criminal Code Ordinance of 1936 that an 'exercise of will' is necessary for responsibility. This is the effect of the famous case of *Mandelbrot* v. *Attorney General*[1] and the subsequent cases which have embedded Agranat J's construction of s. 11 in Israeli law. English lawyers though they may admire this bold step cannot use as an escape route from the confines of the M'Naghten rules the similar doctrine that for any criminal liability there must be a 'voluntary act' which many authorities have said is a fundamental requirement of English criminal law. For this doctrine has always been understood merely to exclude cases where the muscular movements are involuntary as in sleepwalking or 'automatism' or reflex action.[2] Nonetheless there have been changes in England; after a period of frozen immobility the hardened mass of our substantive criminal law is at points softening and yielding to its critics. But both the recent changes and the current criticisms of the law in this matter of criminal

[1] (1956) 10 P.D. 281.

[2] See Edwards, 'Automatism and Criminal Responsibility' 21 *M.L.R.* (1958), p. 375, and Acts of Will and Responsibility, Chap. IV, *supra*. The doctrine as now formulated descends from Austin, *Lectures in Jurisprudence*, Lectures XVIII and XIX.

responsibility have taken a different direction from development in Israel and for this reason may be of some interest to Israeli lawyers.

Let me first say something quite general and very elementary about the historical background to these recent changes. In all advanced legal systems liability to conviction for serious crimes is made dependent, not only on the offender having done those outward acts which the law forbids, but on his having done them in a certain frame of mind or with a certain will. These are the mental conditions or 'mental elements' in criminal responsibility and, in spite of much variation in detail and terminology, they are broadly similar in most legal systems. Even if you kill a man, this is not punishable as murder in most civilised jurisdictions if you do it unintentionally, accidentally or by mistake, or while suffering from certain forms of mental abnormality. Lawyers of the Anglo-American tradition use the Latin phrase *mens rea* (a guilty mind) as a comprehensive name for these necessary mental elements; and according to conventional ideas *mens rea* is a necessary element in liability to be established *before* a verdict. It is not something which is merely to be taken into consideration in determining the sentence or disposal of the convicted person, though it may also be considered for that purpose as well.

I have said that my topic in this lecture is the recent changes in England on this matter, but I shall be concerned less with changes in the law itself than with changes among critics of the law towards the whole doctrine of the mental element in responsibility. This change in critical attitude is, I believe, more important than any particular change in the detail of the doctrine of *mens rea*. I say this because for a century at least most liberal minded people have agreed in treating respect for the doctrine of *mens rea* as a hall-mark of a civilised legal system. Until recently the great aim of most critics of the criminal law has been to secure that the law should take this doctrine very seriously and whole-heartedly. Critics have sought its expansion, and urged that the Courts should be required always to make genuine efforts, when a person is accused of crime, to determine before convicting him whether that person actually did have the knowledge or intention or the sanity or any other mental element which the law, in its definition of crimes, makes a necessary condition of criminal liability. It is true that English law has often wavered on this matter and has even quite recently flirted with the idea that it cannot really afford to inquire into an individual's actual mental state before punishing him. There have always been English judges in whom a remark made in 1477 by Chief Justice Brian of the Common Pleas strikes a sympathetic chord. He said 'The thought of man is not triable; the devil alone knoweth the thought of man'.[3] So there are in English law many compromises on this matter of the relevance of a man's mind to the criminality of his deeds. Not only are there certain crimes of 'strict' liability where neither knowledge, nor negligence is required for conviction, but there are also certain doctrines of 'objective' liability such as was endorsed by the House of Lords in the much criticized case of *The Director of Public Prosecutions* v. *Smith*[4] on which Lord Denning lectured to you three years ago[5]. This doctrine enables a court to impute to an accused person knowledge or an intention which he may not really have had, but which an average man would have had. Theories have been developed in support of this doctrine of 'objective liability' of which the most famous is that expounded by the great American judge, Oliver Wendell Holmes in his book *The Common Law*. Nonetheless generations of progressive minded lawyers and liberal critics of the law have thought of the doctrine of *mens rea* as something to be cherished and extended, and against the scepticism of Chief Justice Brian they could quote the robust assertion of the nineteenth-century Lord Justice Bowen that 'the state of a man's mind is as much a fact as the state of his digestion'.[6] And they would have added that for the criminal law the former was a good deal more important than the latter.

But recently in England progressive and liberal criticism of the law has changed its direction. Though I think this change

[3] *Year Book*, 17 Pasch Ed. IV. f.1. pl. 2. [4] (1961) A.C. 290.
[5] Denning, *Responsibility before the Law*, Jerusalem, 1961.
[6] *Edgington* v. *Fitzmaurice* (1885), 29 Ch. D. 459.

must in the end involve the whole doctrine of *mens rea* it at present mainly concerns the criminal responsibility of mentally abnormal persons, and I can best convey its character by sketching the course taken in the criticism of the law in this matter. The main doctrine of English law until recently was of course the famous M'Naghten Rules formulated by the Judges of the House of Lords in 1843. As everybody knows, according to this doctrine, mental abnormality sufficient to constitute a defence to a criminal charge must consist of three elements: first, the accused, at the time of his act, must have suffered from a defect of reason; secondly, this must have arisen from disease of the mind; thirdly, the result of it must have been that the accused did not know the nature of his act or that it was illegal. From the start English critics denounced these rules because their effect is to excuse from criminal responsibility only those whose mental abnormality resulted in lack of knowledge: in the eyes of these critics this amounted to a dogmatic refusal to acknowledge the fact that a man might know what he was doing and that it was wrong or illegal and yet because of his abnormal mental state might lack the capacity to control his action. This lack of capacity, the critics urged, must be the fundamental point in any intelligible doctrine of responsibility. The point just is that in a civilized system only those who *could have* kept the law should be punished. Why else should we bother about a man's knowledge or intention or other mental element except as throwing light on this?

Angrily and enviously, many of the critics pointed to foreign legal systems which were free of the English obsession with this single element of knowledge as the sole constituent of responsibility. As far back as 1810 the French Code simply excused those suffering from madness (démence) without specifying any particular connexion between this and the particular act done. The German Code of 1871 spoke of inability or impaired ability to recognize the wrongness of conduct or to act in accordance with this recognition. It thus, correctly, according to the critics, treated as crucial to the issue of responsibility not knowledge but the capacity to conform to law. The Belgian Loi de Défence Sociale of 1930 makes no reference to knowledge or intelligence but speaks simply of a person's lack of ability as a consequence of mental abnormality to control his action. So till recently the great aim of the critics inspired by these foreign models was essentially to secure an amendment of the English doctrine of *mens rea* on this point: to supplement its purely cognitive test by a volitional one, admitting that a man might, while knowing that he was breaking the law, be unable to conform to it.

This dispute raged through the nineteenth century and was certainly marked by some curious features. In James Fitzjames Stephen's great *History of the Criminal Law*[7] the dispute is vividly presented as one between doctors and lawyers. The doctors are pictured as accusing the lawyers of claiming to decide a medical or scientific issue about responsibility by out-of-date criteria when they limited legal inquiry to the question of knowledge. The lawyers replied that the doctors, in seeking to give evidence about other matters, were attempting illicitly to thrust upon juries their views on what should excuse a man when charged with a crime: illicitly, because responsibility is a question not of science but of law. Plainly, the argument was here entrapped in the ambiguities of the word 'responsibility' about which more should have been said. But it is also remarkable that in the course of this long dispute no clear statements were made of the reason why the law should recognise any form of insanity as an excuse. The basic question as to what was at stake in the doctrine of *mens rea* was hardly faced. Is it necessary because punishment is conceived of as paying back moral evil done with some essentially retributive 'fitting' equivalent in pain? If so, what state of mind does a theory of retribution require a person punished to have had? Or is a doctrine of *mens rea* necessary because punishment is conceived as primarily a deterrent and this purpose would be frustrated or useless if persons were punished who at the time of their crime lacked certain knowledge or ability? Or is the doctrine to give effect not to a retributive theory but to principles of fairness or justice which require that a man should not

[7] Chap. XIX, Vol. II, 'On the Relation of Madness to Crime'.

be punished and so be used for the ends of others unless he had the capacity and a fair opportunity to avoid doing the thing for which he is punished? Certainly Bentham and Blackstone had something to say on these matters of fundamental principle, but they do not figure much in the century-long war which was waged by English reformers, sometimes in a fog, against the M'Naghten Rules. But what was clear in the fog was that neither party thought of calling the whole doctrine of *mens rea* in question. What was sought was merely amendments or additions to it.

Assault after assault on the M'Naghten Rules were beaten off until 1957. It cannot be said that the defenders of the doctrine used any very sharp rapiers in their defence. The good old English bludgeon which has beaten off so many reforms of English criminal law was enough. When Lord Atkin's Committee recommended in 1923 an addition to the M'Naghten Rules to cater for what it termed 'irresistible impulse', it was enough in the debate in the House of Lords[8] for judicial members to prophesy the harm to society which would inevitably flow from the amendment. Not a word was said to meet the point that the laws of many other countries already conformed to the proposal: nothing was said about the United States where a similar modification of the M'Naghten Rules providing for inability to conform to the law's requirement as well as defects in knowledge had been long accepted in several States without disastrous results. But in 1957, largely as a result of the immensely valuable examination of the whole topic by the Royal Commission on Capital Punishment[9] the law was amended, not as recommended by the Commission, but in the form of a curious compromise. This was the introduction of the idea borrowed from Scots law of a plea of diminished responsibility. S. 2 of the Homicide Act of 1957 provides that, on a murder charge, if what it most curiously calls the accused's 'mental responsibility' was 'substantially'

[8] 57 H.L. Deb. 443–76 (1924), 'if this Bill were passed very grave results would follow' (Lord Sumner, p. 459). 'What a door is being opened!' (Lord Hewart, p. 467). 'This would be a very dangerous change to make' (Lord Cave, p. 475).

[9] Cmd. 8932 (1953).

impaired by mental abnormality, he could be convicted, not of murder, but only of manslaughter, carrying a maximum sentence of imprisonment for life. This change in the law was indeed meagre since it concerned only murder; and even here it was but a half-way house, since the accused was not excused from punishment but was to be punished less than the maximum. The change does not excuse from responsibility but mitigates the penalty.

A word or two about the operation of the new plea of diminished responsibility during the last six years is necessary. The judges at first tended to treat it merely as catering for certain cases on the borderlines of the M'Naghten Rules, not as making a major change. Thus Lord Goddard refused to direct the jury that under the new plea the question of capacity to conform to law and not merely the accused's knowledge was relevant.[10] But the present Lord Chief Justice in a remarkable judgment expressly stated that this was so, and a generous interpretation was given to the section so as to include in the phrase 'abnormality of mind' the condition of the psychopath. He said that it was important to consider not only the accused's knowledge but also his ability 'to exercise will power to control physical acts in accordance with rational judgment'.[11] However, the most remarkable feature of six years' experience of this plea is made evident by the statistics: apprehensions that it might lead to large-scale evasions of punishment have been shown to be quite baseless. For since the Homicide Act almost precisely the same percentage—about 47 per cent—of persons charged with murder escaped conviction on the ground of mental abnormality as before. What has happened is that the plea of insanity under the old M'Naghten Rules has virtually been displaced in murder cases by the new plea.[12] Though satisfactory, in that the old fears of reform have not been realized, the plea certainly has its critics and in part the general change in attitude of which I shall speak has been accelerated by it.

[10] *R.* v. *Spriggs* (1958), 1 Q.B. 270.

[11] *R.* v. *Byrne* (1960), 44 Cr.App. Rep. 246.

[12] For the statistics see *Murder: Home Office Research Unit Report*, H.M.S.O. 1961, Table 7, p. 10.

2

I have said that the change made by the introduction of diminished responsibility was both meagre and half-hearted. Nonetheless it marked the end of an era in the criticism of the law concerning the criminal responsibility of the mentally abnormal. From this point on criticism has largely changed its character. Instead of demanding that the court should take more seriously the task of dividing law-breakers into two classes—those fully responsible and justly punishable because they had an unimpaired capacity to conform to the law, and those who were to be excused for lack of this—critics have come to think this a mistaken approach. Instead of seeking an expansion of the doctrine of *mens rea* they have argued that it should be eliminated and have welcomed the proliferation of offences of strict liability as a step in the right direction and a model for the future. The bolder of them have talked of the need to 'by-pass' or 'dispense with' questions of responsibility and have condemned the old efforts to widen the scope of the M'Naghten Rules as waste of time or worse. Indeed, their attitude to such reforms is like that of the Communist who condemns private charity in a capitalist system because it tends to hide the radical errors of the system and thus preserve it. By far the best informed, most trenchant and influential advocate of these new ideas is Lady Wootton whose powerful work on the subject of criminal responsibility has done much to change and, in my opinion, to raise, the whole level of discussion.[13]

Hence, since 1957 a new scepticism going far beyond the old criticisms has developed. It is indeed a scepticism of the whole institution of criminal punishment so far as it contains elements which differentiate it from a system of purely forward-looking social hygiene in which our only concern, when we have an offender to deal with, is with the future and the rational aims of the prevention of further crime, the protection of society

[13] See her *Social Science and Social Pathology* (1959) esp. Chapter VIII on 'Mental Disorder and the Problem of Moral and Criminal Responsibility'; 'Diminished Responsibility: A Layman's View' 76 *L.Q.R.* (1960), p. 224; *Crime and the Criminal Law* (1963).

and the care and if possible the cure of the offender. For criminal punishment, as even the most progressive older critics of the M'Naghten Rules conceived of it, is *not* mere social hygiene. It differs from such a purely forward-looking system in the stress that it places on something in the past: the state of mind of the accused at the time, not of his trial, but when he broke the law.

To many modern critics this backward-looking reference to the accused's past state of mind as a condition of his liability to compulsory measures seems a useless deflection from the proper forward-looking aims of a rational system of social control. The past they urge is over and done with, and the offender's past state of mind is only important as a diagnosis of the causes of his offence and a prognosis of what can be done now to counter these causes. Nothing in the past, according to this newer outlook, can in itself justify or be required to license what we do to the offender now; that is something to be determined exclusively by reference to the consequences to society and to him. Lady Wootton argues that if the aim of the criminal law is to be the prevention of 'socially damaging actions' not retribution for past wickedness, the conventional doctrine puts *mens rea* 'into the wrong place'.[14] *Mens rea* is on her view relevant only *after* conviction as a guide to what measures should be taken to prevent a recurrence of the forbidden act. She considers it 'illogical' if the aim of the criminal law is prevention to make *mens rea* part of the definition of a crime and a necessary condition of the offender's liability to compulsory measures.[15]

This way of thinking leads to a radical revision of the penal system which in crude outline and in its most extreme form is as follows: Once it has been proved in a court that a person's outward conduct fits the legal definition of some crime, this without proof of any *mens rea*, is sufficient to bring him within the scope of compulsory measures. These may be either of a

[14] See *Crime and the Criminal Law*, p. 52. But she does not consider explicitly whether, even if the aim of the criminal law is to prevent crime, there are not moral objections to applying its sanctions even as preventives to those who lacked the capacity to conform to the Law. See *infra*, pp. 207–8.

[15] Op. cit., p. 51.

penal or therapeutic kind or both; or it may be found that no measures are necessary in a particular case and the offender may be discharged. But the choice between these alternatives is not to be made by reference to the offender's past mental state—his culpability—but by consideration of what steps, in view of his present mental state and his general situation, are likely to have the best consequences for him and for society.

I have called this the extreme form of the new approach because as I have formulated it it is generally applicable to all offenders alike. It is not a system reserved solely for those who could be classed as mentally abnormal. The whole doctrine of *mens rea* would on this extreme version of the theory be dropped from the law; so that the distinctions which at present we draw and think vital to draw before convicting an offender, between, for example, intentional and unintentional wrong-doing, would no longer be relevant at this stage. To show that you have struck or wounded another unintentionally or without negligence would not save you from conviction and liability to such treatment, penal or therapeutic, as the court might deem advisable on evidence of your mental state and character.

This is, as I say, the extreme form of the theory, and it is the form that Lady Wootton now advances.[16] But certainly a less extreme though more complex form is conceivable which would replace, not the whole doctrine of *mens rea*, but only that part of it which concerns the legal responsibility of the mentally abnormal. In this more moderate form of the theory a mentally normal person would still escape conviction if he acted unintentionally or without some other requisite mental element forming part of the definition of the crime charged. The innovation would be that no form of insanity or mental abnormality would bar a conviction, and this would no longer be investigated before conviction.[17] It would be something to be investigated only after conviction to determine what

[16] In *Crime and the Criminal Law* she makes it clear that the elimination or 'withering away' of *mens rea* as a condition of liability is to apply to all its elements not merely to its provision for mental abnormality. Hence strict liability is welcomed as the model for the future (op. cit., pp. 46–57).

[17] Save as indicated *infra* p. 205, n. 31.

measures of punishment or treatment would be most efficacious in the particular case. It is important to observe that most advocates of the elimination of responsibility have been mainly concerned with the inadequacies or absurdities of the existing law in relation to mentally abnormal offenders, and some of these advocates may have intended only the more moderate form of the theory which is limited to such offenders. But I doubt if this is at all representative, for many, including Lady Wootton, have said that no satisfactory line can be drawn between the mentally normal and abnormal offenders: there simply are no clear or reliable criteria. They insist that general definitions of mental health are too vague and too conflicting; we should be freed from all such illusory classifications to treat, in the most appropriate way from the point of view of society, all persons who have actually manifested the behaviour which is the *actus reus* of a crime.[18] The fact that harm was done unintentionally should not preclude an investigation of what steps if any are desirable to prevent a repetition. This scepticism of the possibility of drawing lines between the normal and abnormal offenders commits advocates of the elimination of responsibility to the extreme form of the theory.

Such then are the essentials of the new idea. Of course the phrase 'eliminating responsibility' does sound very alarming and when Lady Wootton's work first made it a centre of discussion the columns of *The Times* newspaper showed how fluttered legal and other dovecotes were. But part at least of the alarm was unnecessary because it arose from the ambiguities of the word 'responsibility'; and it is, I think, still important to distinguish two of the very different things this difficult word may mean. To say that someone is legally responsible for something often means only that under legal rules he is liable to be made either to suffer or to pay compensation in certain eventualities. The expression 'he'll pay for it' covers both these things. In this the primary sense of the word, though a man is normally only responsible for his own actions or the harm he has done, he may be also responsible for the actions of other persons if legal rules so provide. Indeed in this sense

[18] See Wootton, op. cit., p. 51.

a baby in arms or a totally insane person might be legally responsible—again, if the rules so provide; for the word simply means liable to be made to account or pay and we might call this sense of the word 'legal accountability'. But the new idea —the programme of eliminating responsibility—is not, as some have feared, meant to eliminate legal accountability: persons who break the law are not just to be left free. What is to be eliminated are enquiries as to whether a person who has done what the law forbids was responsible at the time he did it and responsible in this sense does not refer to the legal status of accountability. It means the capacity, so far as this is a matter of a man's mind or will, which normal people have to control their actions and conform to law. In this sense of responsibility a man's responsibility can be said to be 'impaired'. That is indeed the language of s. 2 of the Homicide Act 1957 which introduced into English law the idea of diminished responsibility: it speaks of a person's '*mental*' responsibility and in the rubric to s. 2 even of persons 'suffering from' diminished responsibility. It is of course easy to see why this second sense of responsibility (which might be called 'personal responsibility') has grown up alongside the primary idea of legal accountability. It is no doubt because the law normally, though not always, confines legal accountability to persons who are believed to have normal capacities of control.

So perhaps the new ideas are less alarming than they seem at first. They are also less new, and those who advocate them have always been able to point to earlier developments within English law which seem to foreshadow these apparently revolutionary ideas. Lady Wootton herself makes much of the fact that the doctrine of *mens rea* in the case of normal offenders has been watered down by the introduction of strict liability and she deprecates the alarm this has raised. But apart from this, the Courts have often been able to deal with mentally abnormal persons accused of crime without confronting the issue of their personal responsibility at the time of their offence. There are in fact several different ways in which this question may be avoided. A man may be held on account of his mental state to be unfit to plead when brought to trial; or he may be certified insane before trial; or, except on a charge of murder, an accused person might enter a plea of guilty with the suggestion that he should be put on probation with a condition of mental treatment.[19] In fact, only a very small percentage of the mentally abnormal have been dealt with under the M'Naghten Rules, a fact which is understandable since a successful plea under those Rules means detention in Broadmoor for an indefinite period and many would rather face conviction and imprisonment and so may not raise the question of mental abnormality at all. So the old idea of treating mental abnormality as bearing on the question of the accused's responsibility and to be settled before conviction, has with few exceptions only been a reality in murder cases to which alone is the plea of diminished responsibility applicable.

But the most important departure from received ideas incorporated in the doctrine of *mens rea* is the Mental Health Act, 1959, which expands certain principles of older legislation. S. 60 of this Act provides that in any case, except where the crime is not punishable by imprisonment or the sentence is fixed by the law (and this latter exception virtually excludes only murder), the courts may, after conviction of the offender, if two doctors agree that the accused falls into any of four specified categories of mental disorder, order his detention for medical treatment instead of passing a penal sentence, though it requires evidence that such detention is warranted. The four categories of mental disorder are very wide and include even psychopathic disorder in spite of the general lack of clear or agreed criteria of this condition. The courts are told by the statute that in exercising their choice between penal or medical measures to have regard to the nature of the offence and the character and antecedents of the offender. These powers have come to be widely used[20] and are available even in cases where a murder charge has been reduced to manslaughter on a plea of provocation or diminished responsibility.

[19] In 1962 the number of persons over 17 treated in these ways were respectively 36 (unfit to plead), 5 (insane before trial), and 836 (probation with mental treatment). See *Criminal Statistics* 1962.

[20] In 1962 hospital orders under this section were made in respect of 1187 convicted persons (*Criminal Statistics* 1962).

Advocates of the programme of eliminating responsibility welcome the powers given by the Mental Health Act to substitute compulsory treatment for punishment, but necessarily they view it as a compromise falling short of what is required, and we shall understand their own views better if we see why they think so. It falls short in four respects. First the power given to courts to order compulsory treatment instead of punishment is discretionary, and even if the appropriate medical evidence is forthcoming the courts may still administer conventional punishment if they choose. The judges *may* still think in terms of responsibility, and it is plain that they occasionally do so in these cases. Thus in the majority of cases of conviction for manslaughter following on a successful plea of diminished responsibility, the courts have imposed sentences of imprisonment notwithstanding their powers under s. 60 of the Mental Health Act, and the Lord Chief Justice has said that in such cases the prisoner may on the facts be shown to have *some* responsibility for which he must be punished.[21] Secondly, the law itself still preserves a conception of penal methods, such as imprisonment, coloured by the idea that it is a payment for past wickedness and not just an alternative to medical treatment; for though the courts may order medical treatment or punish, they cannot combine these. This of course is a refusal to think, as the new critics demand we should think,[22] of punitive and medical measures as merely different forms of social hygiene to be used according to a prognosis of their effects on the convicted person. Thirdly, as it stands at present, the scheme presupposes that a satisfactory distinction can be drawn on the basis of its four categories of mental disorder between those who are mentally abnormal and those who are not. But the more radical reformers are not merely sceptical about the adequacy of the criteria which distinguish, for example, the psychopath from the normal offender: they would contend that there may exist propensities to certain types of socially harmful behaviour in people who are in other ways not abnormal and that a rational system should attend to these cases.

[21] *R.* v. *Morris* (1961) 45 Cr. App. Rep. 185.
[22] See Wootton, op. cit., pp. 79–80.

But fourthly, and this is most important, the scheme is vitiated for these critics because the courts' powers are only exercisable after the conviction of an offender and, for this conviction, proof of *mens rea* at the time of his offence is still required: the question of the accused's mental abnormality may still be raised before conviction as a defence if the accused so wishes. So the Mental Health Act does not 'by-pass' the whole question of responsibility: it does not eliminate the doctrine of *mens rea*. It expands the courts' discretion in dealing with a convicted person, enabling it to choose between penal and therapeutic measures and making this choice in practice largely independent of the offender's state of mind at the time of his offence. Its great merit is that the mentally abnormal offender who would before have submitted to a sentence of imprisonment rather than raise a plea of insanity under the M'Naghten Rules (because success would mean indeterminate detention in Broadmoor) may now be encouraged to bring forward his mental condition after conviction, in the hope of obtaining a hospital order rather than a sentence of imprisonment.

The question which now awaits our consideration is the merits of the claim that we should proceed from such a system as we now have under the Mental Health Act to one in which the criminal courts were freed altogether from the doctrine of *mens rea* and could proceed to the use of either penal or medical measures at discretion simply on proof that the accused had done the outward acts of a crime. Prisons and hospitals under such a scheme will alike 'be simply "places of safety" in which offenders receive the treatment which experience suggests is most likely to evoke the desired response'.[23]

The case for adopting these new ideas in their entirety has been supported by arguments of varying kinds and quality, and it is very necessary to sift the wheat from the chaff. The weakest of the arguments is perhaps the one most frequently heard, namely, that our concern with personal responsibility incorporated in the doctrine of *mens rea* only makes sense if we subscribe to a retributive theory of punishment according to

[23] Wootton, op. cit., pp. 79–80.

which punishment is used and justified as an 'appropriate' or 'fitting' return for past wickedness and not merely as a preventive of anti-social conduct. This, as I have argued elsewhere,[24] is a philosophical confusion and Lady Wootton falls a victim to it because she makes too crude a dichotomy between 'punishment' and 'prevention'. She does not even mention a moral outlook on punishment which is surely very common, very simple and except perhaps for the determinist perfectly defensible. This is the view that out of considerations of fairness or justice to individuals we should restrict even punishment designed as a 'preventive' to those who had a normal capacity and a fair opportunity to obey. This is still an intelligible ideal of justice to the individuals whom we punish even if we punish them to protect society from the harm that crime does and not to pay back the harm that they have done. And it remains intelligible even if in securing this form of fairness to those whom we punish we secure a lesser measure of conformity to law than a system of total strict liability which repudiated the doctrine of *mens rea*.

But of course it is certainly arguable that, at present, in certain cases, in the application of the doctrine of *mens rea*, we recognize this principle of justice in a way which plays too high a price in terms of social security. For there are indeed cases where the application of *mens rea* operates in surprising and possibly dangerous ways. A man may cause very great harm, may even kill another person, and under the present law neither be punished for it nor subjected to any compulsory medical treatment or supervision. This happened, for example, in February 1961 when a United States Air Force sergeant,[25] after a drunken party, killed a girl, according to his own story, in his sleep. He was tried for murder but the jury were not persuaded by the prosecution, on whom the legal burden of proof rests, that the sergeant's story was false and he was accordingly acquitted and discharged altogether. It is worth observing that in recent years in cases of dangerous driving where the accused claims that he suffered from 'automatism'

[24] 'Punishment and the Elimination of Responsibility', Chap. VII, *supra*.

[25] *The Times*, 18 February 1961 (Staff Sergeant Boshears).

or a sudden lapse of consciousness, the courts have striven very hard to narrow the scope of this defence because of the obvious social dangers of an acquittal of such persons, unaccompanied by any order for compulsory treatment. They have produced a most complex body of law distinguishing between 'sane' and 'insane' automatism each with their special burdens of proof.[26] No doubt such dangerous cases are not very numerous and the risk of their occurrence is one which many people might prefer to run rather than introduce a new system dispensing altogether with proof of *mens rea*. In any case something less extreme than the new system might deal with such cases; for the courts could be given powers in the case of such physically harmful offences to order, notwithstanding an acquittal, any kind of medical treatment or supervision that seemed appropriate.

But the most important arguments in favour of the more radical system in which proof of the outward act alone is enough to make the accused liable to compulsory measures of treatment or punishment, comes from those who, like Lady Wootton, have closely scrutinized the actual working of the old plea of insanity and the plea of diminished responsibility introduced in 1957 by the Homicide Act into cases of homicide. The latter treats mental abnormality as an aspect of *mens rea* and forces the Courts before the verdict to decide the question whether the accused's 'mental responsibility', that is, his capacity to control his actions was 'substantially impaired' at the time of his offence when he killed another person. The conclusion drawn by Lady Wootton from her impressive and detailed study of all the cases (199 in number) in which this plea was raised down to mid-September of 1962, is that this question which is thus forced upon the Courts should be discarded as unanswerable. Here indeed she echoes the cry, often in earlier years thundered from the Bench, that it is impossible to distinguish between an irresistible impulse and an impulse which was merely not resisted by the accused.

But here too if we are to form a balanced view we must distinguish between dubious philosophical contentions and

[26] See *Bratty* v. *Att. Gen. for Northern Ireland* (1961), 3 All E.R., 523 and Cross, 'Reflections on Bratty's Case' 78 *L.Q.R.* (1962), p. 236.

some very good sense. The philosophical arguments (which I will not discuss here in detail) pitch the case altogether too high: they are supposed to show that the question whether a man could have acted differently is *in principle unanswerable* and not merely that in Law Courts we do not usually have clear enough evidence to answer it. Lady Wootton says that a man's responsibility or capacity to resist temptation is something 'buried in [his] consciousness, into which no human being can enter',[27] known if at all only to him and to God: it is not something which other men may ever know; and since 'it is not possible to get inside another man's skin'[28] it is not something of which they can ever form even a reasonable estimate as a matter of probability. Yet strangely enough she does not take the same view of the question which arises under the M'Naghten Rules whether a man knew what he was doing or that it was illegal, although a man's knowledge is surely as much, or as little, locked in his breast as his capacity for self control. Questions about the latter indeed may often be more difficult to answer than questions about a man's knowledge; yet in favourable circumstances if we know a man well and can trust what he says about his efforts or struggles to control himself we may have as good ground for saying 'Well he just could not do it though he tried' as we have for saying 'He didn't know that the pistol was loaded'. And we sometimes may have good general evidence that in certain conditions, e.g. infancy or a clinically definable state, such as depression after childbirth, human beings are unable or less able than the normal adult to master certain impulses. We are not forced by the facts to say of a child or mental defective, who has struggled vainly with tears, merely 'he usually cries when that happens'. We say—and why not?—'he could not stop himself crying though he tried as hard as he could'.

It must however be conceded that such clear cases are very untypical of those that face the Courts where an accused person is often fighting for his life or freedom. Lady Wootton's best arguments are certainly independent of her more debatable

[27] See 'Diminished Responsibility: A Layman's view' 76 *L.Q.R.* (1960), p. 232.
[28] See *Crime and the Criminal Law*, p. 74.

philosophical points about our ability to know what is locked in another's mind or breast. Her central point is that the evidence put before Courts on the question whether the accused lacked the capacity to conform to the law, or whether it was substantially impaired, at the best only shows the *propensity* of the accused to commit crimes of certain sorts. From this, she claims, it is a fallacy to infer that he could not have done otherwise than commit the crime of which he is accused. She calls this fallacious argument 'circular': we infer the accused's lack of capacity to control his actions from his propensity to commit crimes and then both explain this propensity and excuse his crimes by his lack of capacity. Lady Wootton's critics have challenged this view of the medical and other evidence on which the Courts act in these cases.[29] They would admit that it is at any rate in part *through* studying a man's crimes that we may discern his incapacity to control his actions. Nonetheless the evidence for this conclusion is not merely the bare fact that he committed these crimes repeatedly, but the manner and the circumstances and the psychological state in which he did this. Secondly in forming any conclusion about a man's ability to control his action much more than his repeated crimes are taken into account. Anti-social behaviour is not just used to explain and excuse itself, even in the case of the psychopath, the definition of whose disorder presents great problems. I think there is much in these criticisms. Nonetheless the forensic debate before judge and jury of the question whether a mentally disordered person could have controlled his action or whether his capacity to do this was or was not 'substantially impaired' seems to me very often very unreal. The evidence tendered is not only often conflicting, but seems to relate to the specific issue of the accused's power or capacity for control on a specific past occasion only very remotely. I can scarcely believe that on this, the supposed issue, anything coherent penetrates to the minds of the jury after they have heard the difficult expert evidence and heard the judge's warning that these matters are 'incapable of

[29] See N. Walker, 'M'Naghten's Ghost', *The Listener*, 29 Aug. 1963, p. 303.

scientific proof'.[30] And I sympathize with the judges in their difficult task of instructing juries on this plea. In Israel there are no juries to be instructed and the judges themselves must confront these same difficulties in deciding in accordance with the principle of the *Mandelbrot* case whether or not the action of a mentally abnormal person who knew what he was doing occurred 'independently of the exercise of his will'.

Because of these difficulties I would prefer to the present law the scheme which I have termed the 'moderate' form of the new doctrine. Under this scheme *mens rea* would continue to be a necessary condition of liability to be investigated and settled before conviction except so far as it relates to mental abnormality. The innovation would be that an accused person would no longer be able to adduce any form of mental abnormality as a bar to conviction. The question of his mental abnormality would under this scheme be investigated only after conviction and would be primarily concerned with his present rather than his past mental state. His past mental state at the time of his crime would only be relevant so far as it provided ancillary evidence of the nature of his abnormality and indicated the appropriate treatment. This position could perhaps be fairly easily reached by eliminating the pleas of insanity and diminished responsibility and extending the provisions of the Mental Health Act, 1959 to all offences including murder. But I would further provide that in cases where the appropriate direct evidence of mental disorder was forthcoming the Courts should no longer be permitted to think in terms of responsibility and mete out penal sentences instead of compulsory medical treatment. Yet even this moderate reform certainly raises some difficult questions requiring careful consideration.[31]

[30] Per Parker C. J. in *R.* v. *Byrne* (1960) 44 Cr. App. 246 at 258.

[31] Of these difficult questions the following seem the most important.

(1) If the post-conviction inquiry into the convicted person's mental abnormality is to focus on his present state, what should a court do with an offender (a) who suffered some mental disorder at the time of his crime but has since recovered? (b) who was 'normal' at the time of the crime but at the time of his conviction suffers from mental disorder?

(2) The Mental Health Act does not by its terms require the court to be satisfied before making a hospital order that there was any causal connexion between

Many I think would wish to go further than this 'moderate' scheme and would join Lady Wootton in a demand for the elimination of the whole doctrine of *mens rea* or at least in the hope that it will 'wither away'. My reasons for not joining them consist of misgivings on three principal points. The first concerns individual freedom. In a system in which proof of *mens rea* is no longer a necessary condition for conviction, the occasions for official interferences with our lives and for compulsion will be vastly increased. Take, for example, the notion of a criminal assault. If the doctrine of *mens rea* were swept away, every blow, even if it was apparent to a policeman that it was purely accidental or merely careless and therefore not, according to the present law, a criminal assault, would be a matter for investigation under the new scheme, since the possibilities of a curable or treatable condition would have to be investigated and the condition if serious treated by medical or penal methods. No doubt under the new dispensation, as at present, prosecuting authorities would use their common sense; but very considerable discretionary powers would have to be entrusted to them to sift from the mass the cases worth investigation as possible candidates for thereapeutic or penal treatment. No one could view this kind of expansion of police powers with equanimity, for with it will come great uncertainty for the individual: official interferences with his life will be more frequent but he will be less able to predict their incidence if any accidental or careless blow may be an occasion for them.

My second misgiving concerns the idea to which Lady

---

the accused's disorder and his offence, but only provides that the court in the exercise of its discretion shall have regard to the nature of the offence. Would this still be satisfactory if the Courts were bound to make a hospital order if the medical evidence of abnormality is forthcoming?.

(3) The various elements of *mens rea* (knowledge, intention, and the minimum control of muscular movements required for an act) may be absent either in a person otherwise normal or may be absent because of some mental disorder (compare the distinctions now drawn between 'sane' and 'insane' automatism). (See *supra*, p. 202). Presumably it would be desirable that in the latter case there should not be an acquittal; but to identify such cases where there were grounds for suspecting mental abnormality, some investigation of mental abnormality would be necessary before the verdict.

Wootton attaches great importance: that what we now call punishment (imprisonment and the like) and compulsory medical treatment should be regarded just as alternative forms of social hygiene to be used according to the best estimate of their future effects, and no judgment of responsibility should be required before we apply to a convicted person those measures, such as imprisonment, which we now think of as penal. Lady Wootton thinks this will present no difficulty as long as we take a firm hold of the idea that the purpose and justification of the criminal law is to prevent crime and not to pay back criminals for their wickedness. But I do not think objections to detaching the use of penal methods from judgments of responsibility can be disposed of so easily. Though Lady Wootton looks forward to the day when the 'formal distinction' between hospitals and prisons will have disappeared, she does not suggest that we should give up the use of measures such as imprisonment. She contemplates that 'those for whom medicine has nothing to offer'[32] may be sentenced to 'places of safety' to receive 'the treatment which experience suggests is most likely to evoke the desired responses', and though it will only be for the purposes of convenience that their 'places of safety' will be separate from those for whom medicine has something to offer, she certainly accepts the idea that imprisonment may be used for its deterrent effect on the person sentenced to it.

This vision of the future evokes from me two different responses: one is a moral objection and the other a sociological or criminological doubt. The moral objection is this: If we imprison a man who has broken the law in order to deter him and by his example others, we are using him for the benefit of society, and for many people, including myself, this is a step which requires to be justified by (*inter alia*) the demonstration that the person so treated could have helped doing what he did. The individual according to this outlook, which is surely neither esoteric nor confused, has a right not to be used in this way unless he could have avoided doing what he did. Lady Wootton would perhaps dismiss this outlook as a disguised

[32] Op. cit., p. 79–80 ('places of safety' are in quotation marks in her text).

form of a retributive conception of punishment. But it is in fact independent of it as I have attempted to show: for though we must seek a moral licence for punishing a man in his voluntary conduct in breaking the law, the punishment we are then licensed to use may still be directed solely to preventing future crimes on his part or on others' and not to 'retribution'.

To this moral objection it may be replied that it depends wholly on the assumption that imprisonment for deterrent purposes will, under the new scheme, continue to be regarded by people generally as radically distinct from medical treatment and still requiring justification in terms of responsibility. It may be said that this assumption should not be made; for the operation of the system itself will in time cause this distinction to fade, and conviction by a court, followed by a sentence of imprisonment, will in time be assimilated to such experiences as a compulsory medical inspection followed by detention in an isolation hospital. But here my sociological or criminological doubts begin. Surely there are two features which, at present, are among those distinguishing punishment from medical treatment and will have to be stripped away before this assimilation can take place, and the moral objection silenced. One of these is that, unlike medical treatment, we use deterrent punishment to deter not only the individual punished but others by the example of his punishment and the severity of the sentence may be adjusted accordingly. Lady Wootton is very sceptical of the whole notion that we can deter in this way potential offenders and therefore she may be prepared to forego this aspect of punishment altogether. But can we on the present available evidence safely adopt this course for all crime? The second feature distinguishing punishment from treatment is that unlike a medical inspection followed by detention in hospital, conviction by a court followed by a sentence of imprisonment is a public act expressing the odium, if not the hostility, of society for those who break the law. As long as these features attach to conviction and a sentence of imprisonment, the moral objection to their use on those who could not have helped doing what they did will remain. On the other hand, if they cease to attach, will not the law have lost an important element in its authority and deterrent force—as important perhaps for some convicted persons as the deterrent force of the actual measures which it administers?

My third misgiving is this. According to Lady Wootton's argument it is a mistake, indeed 'illogical', to introduce a reference to *mens rea* into the definition of an offence. But it seems that a code of criminal law which omitted any reference in the definition of its offences to mental elements could not possibly be satisfactory. For there are some socially harmful activities which are now and should always be treated as criminal offences which can only be identified by reference to intention or some other mental element. Consider the idea of an attempt to commit a crime. It is obviously desirable that persons who attempt to kill or injure or steal, even if they fail, should be brought before courts for punishment or treatment; yet what distinguishes an attempt which fails from an innocent activity is just the fact that it is a step taken with the intention of bringing about some harmful consequence.

I do not consider my misgivings on these three points as necessarily insuperable objections to the programme of eliminating responsibility. For the first of them rests on a judgment of the value of individual liberty as compared with an increase in social security from harmful activities, and with this comparative judgment others may disagree. The second misgiving in part involves a belief about the dependence of the efficacy of the criminal law on the publicity and odium at present attached to conviction and sentence and on deterrence by example; psychological and sociological researches may one day show that this belief is false. The third objection may perhaps be surmounted by some ingenuity or compromise, since there are many important offences to which it does not apply. Nonetheless I am certain that the questions I have raised here should worry advocates of the elimination of responsibility more than they do; and until they have been satisfactorily answered I do not think we should move the whole way into this part of the Brave New World.

IX

POSTSCRIPT: RESPONSIBILITY AND RETRIBUTION

THE essays in this volume are all concerned with the legal doctrine which requires, as a normal condition of liability to punishment, that the person to be punished should, at the time of his offence, have had a certain knowledge or intention, or possessed certain powers of understanding and control. This doctrine prescribing the psychological criteria of responsibility takes different forms in different legal systems, but in all its forms it has presented both problems of analysis and problems of policy and moral justification. It is no easy matter to determine precisely what English law actually requires when it is said to require, or to treat as sufficient for liability, a certain 'intention' or an 'act of will' or 'recklessness' or 'negligence'; hence some of the preceding essays are concerned in part with such problems of analysis. But most of them are also concerned with problems of justification: with the credentials of principles or 'theories of punishment' which require liability to punishment to be restricted by reference to such psychological conditions, and with the claims of newer theories that would eliminate these restrictions either completely or in part. A central theme of these essays is that it is not only within the framework of a retributive theory of punishment that insistence on the importance of these restrictions makes sense; there are important reasons, both moral and prudential, for adhering to these restrictions which are perfectly consistent with a general utilitarian conception of the aim of punishment.

In most of these essays I have attempted to confront these issues without any full-scale discussion of the notions of Responsibility and Retribution, though I turned aside to distinguish, in the first of these essays, two meanings of 'retribution' and, in the last essay, two meanings of 'responsibility'. The distinctions I made there have drawn fire from some critics, and it is plain from the criticism that some more comprehensive account of the complexities and ambiguities of these notions is required. The purpose of this postscript is to supply it.

Part One: Responsibility

I

A wide range of different, though connected, ideas is covered by the expressions 'responsibility', 'responsible', and 'responsible for', as these are standardly used in and out of the law. Though connexions exist between these different ideas, they are often very indirect, and it seems appropriate to speak of different *senses* of these expressions. The following simple story of a drunken sea captain who lost his ship at sea can be told in the terminology of responsibility to illustrate, with stylistically horrible clarity, these differences of sense.

'As captain of the ship, X was responsible for the safety of his passengers and crew. But on his last voyage he got drunk every night and was responsible for the loss of the ship with all aboard. It was rumoured that he was insane, but the doctors considered that he was responsible for his actions. Throughout the voyage he behaved quite irresponsibly, and various incidents in his career showed that he was not a responsible person. He always maintained that the exceptional winter storms were responsible for the loss of the ship, but in the legal proceedings brought against him he was found criminally responsible for his negligent conduct, and in separate civil proceedings he was held legally responsible for the loss of life and property. He is still alive and he is morally responsible for the deaths of many women and children.'

This welter of distinguishable senses of the word 'responsibility' and its grammatical cognates can, I think, be profitably reduced by division and classification. I shall distinguish four heads of classification to which I shall assign the following names:

(*a*) Role-Responsibility
(*b*) Causal-Responsibility
(*c*) Liability-Responsibility
(*d*) Capacity-Responsibility.

I hope that in drawing these dividing lines, and in the exposition which follows, I have avoided the arbitrary pedantries of classificatory systematics, and that my divisions pick out and clarify the main, though not all, varieties of responsibility to which reference is constantly made, explicitly or implicitly, by moralists, lawyers, historians, and ordinary men. I relegate to the notes[1] discussion of what unifies these varieties and explains the extension of the terminology of responsibility.

2. ROLE-RESPONSIBILITY

A sea captain is responsible for the safety of his ship, and that is his responsibility, or one of his responsibilities. A husband is responsible for the maintenance of his wife; parents for the upbringing of their children; a sentry for alerting the guard at the enemy's approach; a clerk for keeping the accounts of his firm. These examples of a person's responsibilities suggest the generalization that, whenever a person occupies a distinctive place or office in a social organization, to which specific duties are attached to provide for the welfare of others or to advance in some specific way the aims or purposes of the organization, he is properly said to be responsible for the performance of these duties, or for doing what is necessary to fulfil them. Such duties are a person's responsibilities. As a guide to this sense of responsibility this generalization is, I think, adequate, but the idea of a distinct role or place or office is, of course, a vague one, and I cannot undertake to make it very precise. Doubts about its extension to marginal cases will always arise. If two friends, out on a mountaineering expedition, agree that the one shall look after the food and the other the maps, then the one is correctly said to be responsible for the food, and the other for the maps, and I would classify this as a case of role-responsibility. Yet such fugitive or temporary assignments with specific duties would not usually be considered

[1] *infra*. pp. 264-5.

by sociologists, who mainly use the word, as an example of a 'role'. So 'role' in my classification is extended to include a task assigned to any person by agreement or otherwise. But it is also important to notice that not all the duties which a man has in virtue of occupying what in a quite strict sense of role is a distinct role, are thought or spoken of as 'responsibilities'. A private soldier has a duty to obey his superior officer and, if commanded by him to form fours or present arms on a given occasion, has a duty to do so. But to form fours or present arms would scarcely be said to be the private's responsibility; nor would he be said to be responsible for doing it. If on the other hand a soldier was ordered to deliver a message to H.Q. or to conduct prisoners to a base camp, he might well be said to be responsible for doing these things, and these things to be his responsibility. I think, though I confess to not being sure, that what distinguishes those duties of a role which are singled out as responsibilities is that they are duties of a relatively complex or extensive kind, defining a 'sphere of responsibility' requiring care and attention over a protracted period of time, while short-lived duties of a very simple kind, to do or not do some specific act on a particular occasion, are not termed responsibilities. Thus a soldier detailed off to keep the camp clean and tidy for the general's visit of inspection has this as his sphere of responsibility and is responsible for it. But if merely told to remove a piece of paper from the approaching general's path, this would be at most his duty.

A 'responsible person', 'behaving responsibly' (not 'irresponsibly'), require for their elucidation a reference to role-responsibility. A responsible person is one who is disposed to take his duties seriously; to think about them, and to make serious efforts to fulfil them. To behave responsibly is to behave as a man would who took his duties in this serious way. Responsibilities in this sense may be either legal or moral, or fall outside this dichotomy. Thus a man may be morally as well as legally responsible for the maintenance of his wife and children, but a host's responsibility for the comfort of his guests, and a referee's responsibility for the control of the players is neither legal nor moral, unless the word 'moral' is unilluminatingly used simply to exclude legal responsibility.

3. CAUSAL RESPONSIBILITY

'The long drought was responsible for the famine in India'. In many contexts, as in this one, it is possible to substitute for the expression 'was responsible for' the words 'caused' or 'produced' or some other causal expression in referring to consequences, results, or outcomes. The converse, however, is not always true. Examples of this causal sense of responsibility are legion. 'His neglect was responsible for her distress.' 'The Prime Minister's speech was responsible for the panic.' 'Disraeli was responsible for the defeat of the Government.' 'The icy condition of the road was responsible for the accident.' The past tense of the verb used in this causal sense of the expression 'responsible for' should be noticed. If it is said of a living person, who has in fact caused some disaster, that he *is* responsible for it, this is not, or not merely, an example of causal responsibility, but of what I term 'liability-responsibility'; it asserts his liability on account of the disaster, even though it is also true that he is responsible in that sense *because* he caused the disaster, and that he caused the disaster may be expressed by saying that he was responsible for it. On the other hand, if it is said of a person no longer living that he was responsible for some disaster, this may be either a simple causal statement or a statement of liability-responsibility, or both.

From the above examples it is clear that in this causal sense not only human beings but also their actions or omissions, and things, conditions, and events, may be said to be responsible for outcomes. It is perhaps true that only where an outcome is thought unfortunate or felicitous is its cause commonly spoken of as responsible for it. But this may not reflect any aspect of the meaning of the expression 'responsible for'; it may only reflect the fact that, except in such cases, it may be pointless and hence rare to pick out the causes of events. It is sometimes suggested that, though we may speak of a human being's action as responsible for some outcome in a purely causal sense, we do not speak of a person, as distinct from his actions, as responsible for an outcome, unless he is felt to deserve censure or praise. This is, I think, a mistake. History books are full of examples to the contrary. 'Disraeli was responsible for the defeat of the Government' need not carry even an implication that he was deserving of censure or praise; it may be purely a statement concerned with the contribution made by one human being to an outcome of importance, and be entirely neutral as to its moral or other merits. The contrary view depends, I think, on the failure to appreciate sufficiently the ambiguity of statements of the form 'X *was* responsible for Y' as distinct from 'X *is* responsible for Y' to which I have drawn attention above. The former expression in the case of a person no longer living may be (though it *need* not be) a statement of liability-responsibility.

4. LEGAL LIABILITY-RESPONSIBILITY

Though it was noted that role-responsibility might take either legal or moral form, it was not found necessary to treat these separately. But in the case of the present topic of liability-responsibility, separate treatment seems advisable. For responsibility seems to have a wider extension in relation to the law than it does in relation to morals, and it is a question to be considered whether this is due merely to the general differences between law and morality, or to some differences in the sense of responsibility involved.

When legal rules require men to act or abstain from action, one who breaks the law is usually liable, according to other legal rules, to punishment for his misdeeds, or to make compensation to persons injured thereby, and very often he is liable to both punishment and enforced compensation. He is thus liable to be 'made to pay' for what he has done in either or both of the senses which the expression 'He'll pay for it' may bear in ordinary usage. But most legal systems go much further than this. A man may be legally punished on account of what his servant has done, even if he in no way caused or instigated or even knew of the servant's action, or knew of the likelihood of his servant so acting. Liability in such circumstances is rare in modern systems of criminal law; but it is common in all systems of civil law for men to be made to pay compensation for injuries caused by others, generally their servants or employees. The law of most countries goes further still. A man may be liable to pay compensation for harm suffered by others, though neither he nor his servants have caused it. This is so, for example, in Anglo-American law when the harm is caused by dangerous things which escape from a man's possession, even if their escape is not due to any act or omission of his or his servants, or if harm is caused to a man's employees by defective machinery whose defective condition he could not have discovered.

It will be observed that the facts referred to in the last paragraph are expressed in terms of 'liability' and not 'responsibility'. In the preceding essay in this volume I ventured the general statement that to say that someone is legally responsible for something often means that under legal rules he is liable to be made either to suffer or to pay compensation in certain eventualities. But I now think that this simple account of liability-responsibility is in need of some considerable modification. Undoubtedly, expressions of the form 'he is legally responsible for Y' (where Y is some action or harm) and 'he is legally liable to be punished or to be made to pay compensation for Y' are very closely connected, and sometimes they are used as if they were identical in meaning. Thus, where one legal writer speaks of 'strict responsibility' and 'vicarious responsibility', another speaks of 'strict liability' and 'vicarious liability'; and even in the work of a single writer the expressions 'vicarious responsibility' and 'vicarious liability' are to be found used without any apparent difference in meaning, implication, or emphasis. Hence, in arguing that it was for the law to determine the mental conditions of responsibility, Fitzjames Stephen claimed that this must be so because 'the meaning of responsibility is liability to punishment'.[2]

But though the abstract expressions 'responsibility' and 'liability' are virtually equivalent in many contexts, the

[2] *A History of The Criminal Law*, Vol. II, p. 183.

statement that a man is responsible for his actions, or for some act or some harm, is usually not identical in meaning with the statement that he is liable to be punished or to be made to pay compensation for the act or the harm, but is directed to a narrower and more specific issue. It is in this respect that my previous account of liability-responsibility needs qualification.

The question whether a man is or is not legally liable to be punished for some action that he has done opens up the quite general issue whether all of the various requirements for criminal liability have been satisfied, and so will include the question whether the kind of action done, whatever mental element accompanied it, was ever punishable by law. But the question whether he is or is not legally responsible for some action or some harm is usually not concerned with this general issue, but with the narrower issue whether any of a certain range of conditions (mainly, but not exclusively, psychological) are satisfied, it being assumed that all other conditions are satisfied. Because of this difference in scope between questions of liability to punishment and questions of responsibility, it would be somewhat misleading, though not unintelligible, to say of a man who had refused to rescue a baby drowning in a foot of water, that he was not, according to English law, legally responsible for leaving the baby to drown or for the baby's death, if all that is meant is that he was not liable to punishment because refusing aid to those in danger is not generally a crime in English law. Similarly, a book or article entitled 'Criminal Responsibility' would not be expected to contain the whole of the substantive criminal law determining the conditions of liability, but only to be concerned with a specialized range of topics such as mental abnormality, immaturity, *mens rea*, strict and vicarious liability, proximate cause, or other general forms of connexion between acts and harm sufficient for liability. These are the specialized topics which are, in general, thought and spoken of as 'criteria' of responsibility. They may be divided into three classes: (i) mental or psychological conditions; (ii) causal or other forms of connexion between act and harm; (iii) personal relationships rendering one man liable to be punished or to pay for the acts of another. Each of these three classes requires some separate discussion.

(i) *Mental or psychological criteria of responsibility.* In the criminal law the most frequent issue raised by questions of responsibility, as distinct from the wider question of liability, is whether or not an accused person satisfied some mental or psychological condition required for liability, or whether liability was strict or absolute, so that the usual mental or psychological conditions were not required. It is, however, important to notice that these psychological conditions are of two sorts, of which the first is far more closely associated with the use of the word responsibility than the second. On the one hand, the law of most countries requires that the person liable to be punished should at the time of his crime have had the capacity to understand what he is required by law to do or not to do, to deliberate and to decide what to do, and to control his conduct in the light of such decisions. Normal adults are generally assumed to have these capacities, but they may be lacking where there is mental disorder or immaturity, and the possession of these normal capacities is very often signified by the expression 'responsible for his actions'. This is the fourth sense of responsibility which I discuss below under the heading of 'Capacity-Responsibility'. On the other hand, except where responsibility is strict, the law may excuse from punishment persons of normal capacity if, on particular occasions where their outward conduct fits the definition of the crime, some element of intention or knowledge, or some other of the familiar constituents of *mens rea*, was absent, so that the particular action done was defective, though the agent had the normal capacity of understanding and control. Continental codes usually make a firm distinction between these two main types of psychological conditions: questions concerning general capacity are described as matters of responsibility or 'imputability', whereas questions concerning the presence or absence of knowledge or intention on particular occasions are not described as matters of 'imputability', but are referred to the topic of 'fault' (*schuld, faute, dolo*, &c.).

English law and English legal writers do not mark quite so firmly this contrast between general capacity and the knowledge or intention accompanying a particular action; for the expression *mens rea* is now often used to cover all the variety of psychological conditions required for liability by the law, so that both the person who is excused from punishment because of lack of intention or some ordinary accident or mistake on a particular occasion and the person held not to be criminally responsible on account of immaturity or insanity are said not to have the requisite *mens rea*. Yet the distinction thus blurred by the extensive use of the expression *mens rea* between a persistent incapacity and a particular defective action is indirectly marked in terms of responsibility in most Anglo-American legal writing, in the following way. When a person is said to be not responsible for a particular act or crime, or when (as in the formulation of the M'Naghten Rules and s. 2 of the Homicide Act, 1957) he is said not to be responsible for his 'acts and omissions in doing' some action on a particular occasion, the reason for saying this is usually some mental abnormality or disorder. I have not succeeded in finding cases where a normal person, merely lacking some ordinary element of knowledge or intention on a particular occasion, is said for that reason not to be responsible for that particular action, even though he is for that reason not liable to punishment. But though there is this tendency in statements of liability-responsibility to confine the use of the expression 'responsible' and 'not responsible' to questions of mental abnormality or general incapacity, yet all the psychological conditions of liability are to be found discussed by legal writers under such headings as 'Criminal Responsibility' or Principles of Criminal Responsibility'. Accordingly I classify them here as criteria of responsibility. I do so with a clear conscience, since little is to be gained in clarity by a rigid division which the contemporary use of the expression *mens rea* often ignores.

The situation is, however, complicated by a further feature of English legal and non-legal usage. The phrase 'responsible for his actions' is, as I have observed, frequently used to refer to the capacity-responsibility of the normal person, and, so used, refers to one of the major criteria of liability-responsibility. It is so used in s. 2 of the Homicide Act 1957, which speaks of a person's mental 'responsibility' for his actions being *impaired*, and in the rubric to the section, which speaks of persons 'suffering from diminished responsibility'. In this sense the expression is the name or description of a psychological condition. But the expression is also used to signify liability-responsibility itself, that is, liability to punishment so far as such liability depends on psychological conditions, and is so used when the law is said to 'relieve insane persons of responsibility for their actions'. It was probably also so used in the form of verdict returned in cases of successful pleas of insanity under English law until this was altered by the Insanity Act 1964: the verdict was 'guilty but insane so as not to be responsible according to law for his actions'.

(ii) *Causal or other forms of connexion with harm.* Questions of legal liability-responsibility are not limited in their scope to psychological conditions of either of the two sorts distinguished above. Such questions are also (though more frequently in the law of tort than in the criminal law) concerned with the issue whether some form of connexion between a person's act and some harmful outcome is sufficient according to law to make him liable; so if a person is accused of murder the question whether he was or was not legally responsible for the death may be intended to raise the issue whether the death was too remote a consequence of his acts for them to count as its cause. If the law, as frequently in tort, is not that the defendant's action should have caused the harm, but that there be some other form of connexion or relationship between the defendant and the harm, e.g. that it should have been caused by some dangerous thing escaping from the defendant's land, this connexion or relationship is a condition of civil responsibility for harm, and, where it holds, the defendant is said to be legally responsible for the harm. No doubt such questions of connexion with harm are also frequently phrased in terms of liability.

(iii) *Relationship with the agent.* Normally in criminal law the minimum condition required for liability for punishment is that the person to be punished should himself have done what the law forbids, at least so far as outward conduct is concerned; even if liability is 'strict'; it is not enough to render him liable for punishment that someone else should have done it. This is often expressed in the terminology of responsibility (though here, too, 'liability' is frequently used instead of 'responsibility') by saying that, generally, vicarious responsibility is not known to the criminal law. But there are exceptional cases; an innkeeper is liable to punishment if his servants, without his knowledge and against his orders, sell liquor on his premises after hours. In this case he is vicariously responsible for the sale, and of course, in the civil law of tort there are many situations in which a master or employer is liable to pay compensation for the torts of his servant or employee, and is said to be vicariously responsible.

It appears, therefore, that there are diverse types of criteria of legal liability-responsibility: the most prominent consist of certain mental elements, but there are also causal or other connexions between a person and harm, or the presence of some relationship, such as that of master and servant, between different persons. It is natural to ask why these very diverse conditions are singled out as criteria of responsibility, and so are within the scope of questions about responsibility, as distinct from the wider question concerning liability for punishment. I think that the following somewhat Cartesian figure may explain this fact. If we conceive of a person as an embodied mind and will, we may draw a distinction between two questions concerning the conditions of liability and punishment. The first question is what general types of outer conduct (*actus reus*) or what sorts of harm are required for liability? The second question is how closely connected with such conduct or such harm must the embodied mind or will of an individual person be to render him liable to punishment? Or, as some would put it, to what extent must the embodied mind or will be the author of the conduct or the harm in order to render him liable? Is it enough that the person made the appropriate bodily movements? Or is it required that he did so when possessed of a certain capacity of control and with a certain knowledge or intention? Or that he caused the harm or stood in some other relationship to it, or to the actual doer of the deed? The legal rules, or parts of legal rules, that answer these various questions define the various forms of connexion which are adequate for liability, and these constitute conditions of legal responsibility which form only a part of the total conditions of liability for punishment, which also include the definitions of the *actus reus* of the various crimes.

We may therefore summarize this long discussion of legal liability-responsibility by saying that, though in certain general contexts legal responsibility and legal liability have the same meaning, to say that a man is legally responsible for some act or harm is to state that his connexion with the act or harm is sufficient according to law for liability. Because responsibility and liability are distinguishable in this way, it will make sense to say that because a person is legally responsible for some action he is liable to be punished for it.

5. LEGAL LIABILITY RESPONSIBILITY AND MORAL BLAME

My previous account of legal liability-responsibility, in which I claimed that in one important sense to say that a person is legally responsible meant that he was legally liable for punishment or could be made to pay compensation, has been criticized on two scores. Since these criticisms apply equally to the above amended version of my original account, in which I distinguish the general issue of liability from the narrower issue of responsibility, I shall consider these criticisms here. The first criticism, made by Mr. A. W. B. Simpson,[3] insists on the strong connexion between statements of legal responsibility and moral judgment, and claims that even lawyers tend to confine statements that a person is legally responsible for something to cases where he is considered morally blameworthy,

[3] In a review of 'Changing Conceptions of Responsibility', Chap. VIII, *supra*, in *Crim.L.R.* (1966) 124.

and, where this is not so, tend to use the expression 'liability' rather than 'responsibility'. But, though moral blame and legal responsibility may be connected in some ways, it is surely not in this simple way. Against any such view not only is there the frequent use already mentioned of the expressions 'strict responsibility' and 'vicarious responsibility', which are obviously independent of moral blameworthiness, but there is the more important fact that we can, and frequently do, intelligibly debate the question whether a mentally disordered or very young person who has been held legally responsible for a crime is morally blameworthy. The coincidence of legal responsibility with moral blameworthiness may be a laudable ideal, but it is not a necessary truth nor even an accomplished fact.

The suggestion that the statement that a man is responsible generally means that he is blameworthy and not that he is liable to punishment is said to be supported by the fact that it is possible to cite, without redundancy, the fact that a person is responsible as a ground or reason for saying that he is liable to punishment. But, if the various kinds or senses of responsibility are distinguished, it is plain that there are many explanations of this last mentioned fact, which are quite independent of any essential connexion between legal responsibility and moral blameworthiness. Thus cases where the statement that the man is responsible constitutes a reason for saying that he is liable to punishment may be cases of role-responsibility (the master is legally responsible for the safety of his ship, therefore he is liable to punishment if he loses it) or capacity-responsibility (he was responsible for his actions therefore he is liable to punishment for his crimes); or they may even be statements of liability-responsibility, since such statements refer to part only of the conditions of liability and may therefore be given, without redundancy, as a reason for liability to punishment. In any case this criticism may be turned against the suggestion that responsibility is to be equated with moral blameworthiness; for plainly the statement that someone is responsible may be given as part of the reason for saying that he is morally blameworthy.

6. LIABILITY RESPONSIBILITY FOR PARTICULAR ACTIONS

An independent objection is the following, made by Mr. George Pitcher.[4] The wide extension I have claimed for the notion of liability-responsibility permits us to say not only that a man is legally responsible in this sense for the consequences of his action, but also for his action or actions. According to Mr. Pitcher 'this is an improper way of talking', though common amongst philosophers. Mr. Pitcher is concerned primarily with moral, not legal, responsibility, but even in a moral context it is plain that there is a very well established use of the expression 'responsible for his actions' to refer to capacity-responsibility for which Mr. Pitcher makes no allowance. As far as the law is concerned, many examples may be cited from both sides of the Atlantic where a person may be said to be responsible for his actions, or for his act, or for his crime, or for his conduct. Mr. Pitcher gives, as a reason for saying that it is improper to speak of a man being responsible for his own actions, the fact that a man does not produce or cause his own actions. But this argument would prove far too much. It would rule out as improper not only the expression 'responsible for his actions', but also our saying that a man was responsible vicariously or otherwise for harmful outcomes which he had not caused, which is a perfectly well established legal usage.

None the less, there are elements of truth in Mr Pitcher's objection. First, it seems to be the case that even where a man is said to be legally responsible for what he has done, it is rare to find this expressed by a phrase conjoining the verb of action with the expression 'responsible for'. Hence, 'he is legally responsible for killing her' is not usually found, whereas 'he is legally responsible for her death' is common, as are the expressions 'legally responsible for his act (in killing her)'; 'legally responsible for his crime'; or, as in the official formulation of the M'Naghten Rules, 'responsible for his actions or omissions in doing or being a party to the killing'. These

[4] In 'Hart on Action and Responsibility', *The Philosophical Review* (1960), p. 266.

common expressions in which a noun, not a verb, follows the phrase 'responsible for' are grammatically similar to statements of causal responsibility, and the tendency to use the same form no doubt shows how strongly the overtones of causal responsibility influence the terminology ordinarily used to make statements of liability-responsibility. There is, however, also in support of Mr. Pitcher's view, the point already cited that, even in legal writing, where a person is said to be responsible for his act or his conduct, the relevant mental element is usually the question of insanity or immaturity, so that the ground in such cases for the assertion that the person is responsible or is not responsible for his act is the presence of absence of 'responsibility for actions' in the sense of capacity-responsibility, and not merely the presence or absence of knowledge or intention in relation to the particular act.

7. MORAL LIABILITY-RESPONSIBILITY

How far can the account given above of legal liability-responsibility be applied *mutatis mutandis* to moral responsibility? The *mutanda* seem to be the following: 'deserving blame' or 'blameworthy' will have to be substituted for 'liable to punishment', and 'morally bound to make amends or pay compensation' for 'liable to be made to pay compensation'. Then the moral counterpart to the account given of legal liability-responsibility would be the following: to say that a person is morally responsible for something he has done or for some harmful outcome of his own or others' conduct, is to say that he is morally blameworthy, or morally obliged to make amends for the harm, so far as this depends on certain conditions: these conditions relate to the character or extent of a man's control over his own conduct, or to the causal or other connexion between his action and harmful occurrences, or to his relationship with the person who actually did the harm.

In general, such an account of the meaning of 'morally responsible' seems correct, and the striking differences between legal and moral responsibility are due to substantive differences between the content of legal and moral rules and principles rather than to any variation in meaning of responsibility when conjoined with the word 'moral' rather than 'legal'. Thus, both in the legal and the moral case, the criteria of responsibility seem to be restricted to the psychological elements involved in the control of conduct, to causal or other connexions between acts and harm, and to the relationships with the actual doer of misdeeds. The interesting differences between legal and moral responsibility arise from the differences in the particular criteria falling under these general heads. Thus a system of criminal law may make responsibility strict, or even absolute, not even exempting very young children or the grossly insane from punishment; or it may vicariously punish one man for what another has done, even though the former had no control of the latter; or it may punish an individual or make him compensate another for harm which he neither intended nor could have foreseen as likely to arise from his conduct. We may condemn such a legal system which extends strict or vicarious responsibility in these ways as barbarous or unjust, but there are no conceptual barriers to be overcome in speaking of such a system as a legal system, though it is certainly arguable that we should not speak of 'punishment' where liability is vicarious or strict. In the moral case, however, greater conceptual barriers exist: the hypothesis that we might hold individuals morally blameworthy for doing things which they could not have avoided doing, or for things done by others over whom they had no control, conflicts with too many of the central features of the idea of morality to be treated merely as speculation about a rare or inferior kind of moral system. It may be an exaggeration to say that there could not logically be such a morality or that blame administered according to principles of strict or vicarious responsibility, even in a minority of cases, could not logically be moral blame; none the less, admission of such a system as a morality would require a profound modification in our present concept of morality, and there is no similar requirement in the case of law.

Some of the most familiar contexts in which the expression 'responsibility' appears confirm these general parallels between legal and moral liability-responsibility. Thus in the famous question 'Is moral responsibility compatible with determinism?' the expression 'moral responsibility' is apt just because the bogey raised by determinism specifically relates to the usual criteria of responsibility; for it opens the question whether, if 'determinism' were true, the capacities of human beings to control their conduct would still exist or could be regarded as adequate to justify moral blame.

In less abstract or philosophical contexts, where there is a present question of blaming someone for some particular act, the assertion or denial that a person is morally responsible for his actions is common. But this expression is as ambiguous in the moral as in the legal case: it is most frequently used to refer to what I have termed 'capacity-responsibility', which is the most important criterion of moral liability-responsibility; but in some contexts it may also refer to moral liability-responsibility itself. Perhaps the most frequent use in moral contexts of the expression 'responsible for' is in cases where persons are said to be morally responsible for the outcomes or results of morally wrong conduct, although Mr. Pitcher's claim that men are never said in ordinary usage to be responsible for their actions is, as I have attempted to demonstrate above with counter-examples, an exaggerated claim.

8. CAPACITY-RESPONSIBILITY

In most contexts, as I have already stressed, the expression 'he is responsible for his actions' is used to assert that a person has certain normal capacities. These constitute the most important criteria of moral liability-responsibility, though it is characteristic of most legal systems that they have given only a partial or tardy recognition to all these capacities as general criteria of legal responsibility. The capacities in question are those of understanding, reasoning, and control of conduct: the ability to understand what conduct legal rules or morality require, to deliberate and reach decisions concerning these requirements, and to conform to decisions when made. Because 'responsible for his actions' in this sense refers not to a legal status but to certain complex psychological characteristics of persons, a person's responsibility for his actions may intelligibly be said to be 'diminished' or 'impaired' as well as altogether absent, and persons may be said to be 'suffering from diminished responsibility' much as a wounded man may be said to be suffering from a diminished capacity to control the movements of his limbs.

No doubt the most frequent occasions for asserting or denying that a person is 'responsible for his actions' are cases where questions of blame or punishment for particular actions are in issue. But, as with other expressions used to denote criteria of responsibility, this one also may be used where no particular question of blame or punishment is in issue, and it is then used simply to describe a person's psychological condition. Hence it may be said purely by way of description of some harmless inmate of a mental institution, even though there is no present question of his misconduct, that he is a person who is not responsible for his actions. No doubt if there were no social practice of blaming and punishing people for their misdeeds, and excusing them from punishment because they lack the normal capacities of understanding and control, we should lack this shorthand description for describing their condition which we now derive from these social practices. In that case we should have to describe the condition of the inmate directly, by saying that he could not understand what people told him to do, or could not reason about it, or come to, or adhere to any decisions about his conduct.

Legal systems left to themselves may be very niggardly in their admission of the relevance of liability to legal punishment of the several capacities, possession of which are necessary to render a man morally responsible for his actions. So much is evident from the history sketched in the preceding chapter of the painfully slow emancipation of English criminal law from the narrow, cognitive criteria of responsibility formulated in the M'Naghten Rules. Though some continental legal systems have been willing to confront squarely the question whether the accused 'lacked the ability to recognize the wrongness of his conduct and to act in accordance with that recognition,'[5] such an issue, if taken seriously, raises formidable difficulties of proof, especially before juries. For this reason I think that, instead of a close determination of such questions of capacity, the apparently coarser-grained technique of exempting persons from liability to punishment if they fall into certain recognized categories of mental disorder is likely to be increasingly used. Such exemption by general category is a technique long known to English law; for in the case of very young children it has made no attempt to determine, as a condition of liability, the question whether on account of their immaturity they could have understood what the law required and could have conformed to its requirements, or whether their responsibility on account of their immaturity was 'substantially impaired', but exempts them from liability for punishment if under a specified age. It seems likely that exemption by medical category rather than by individualized findings of absent or diminished capacity will be found more likely to lead in practice to satisfactory results, in spite of the difficulties pointed out in the last essay in the discussion of s. 60 of the Mental Health Act, 1959.

Though a legal system may fail to incorporate in its rules any psychological criteria of responsibility, and so may apply its sanction to those who are not morally blameworthy, it is none the less dependent for its efficacy on the possession by a sufficient number of those whose conduct it seeks to control of the capacities of understanding and control of conduct which constitute capacity-responsibility. For if a large proportion of those concerned could not understand what the law required them to do or could not form and keep a decision to obey, no legal system could come into existence or continue to exist. The general possession of such capacities is therefore a condition of the *efficacy* of law, even though it is not made a condition of liability to legal sanctions. The same condition of efficacy attaches to all attempts to regulate or control human conduct by forms of *communication:* such as orders, commands, the invocation of moral or other rules or principles, argument, and advice.

[5] German Criminal Code, Art. 51.

'The notion of prevention through the medium of the mind assumes mental ability adequate to restraint'. This was clearly seen by Bentham and by Austin, who perhaps influenced the seventh report of the Criminal Law Commissioners of 1833 containing this sentence. But they overstressed the point; for they wrongly assumed that this condition of efficacy must also be incorporated in legal rules as a condition of liability. This mistaken assumption is to be found not only in the explanation of the doctrine of *mens rea* given in Bentham's and Austin's works, but is explicit in the Commissioners' statement preceding the sentence quoted above that 'the object of penal law being the prevention of wrong, the principle does not extend to mere involuntary acts or even to harmful consequences the result of inevitable accident'. The case of morality is however different in precisely this respect: the possession by those to whom its injunctions are addressed of 'mental ability adequate to restraint' (capacity-responsibility) has there a double status and importance. It is not only a condition of the efficacy of morality; but a system or practice which did not regard the possession of these capacities as a necessary condition of liability, and so treated blame as appropriate even in the case of those who lacked them, would not, as morality is at present understood, be a morality.

Part Two: Retribution

I

IN the first of these essays I made some attempt to clarify the idea of retribution by distinguishing what I there called Retribution as a General Justifying Aim from retribution in the distribution of punishment. But it is plain enough that I have not done justice to the variety and complexity of this notion, and some rather unrewarding disputes about the morality of punishment continue to flourish, in part at least, because some of its ambiguities are still undetected. So in the effort to bring them to light, I shall explore here some further reaches of the subject.

One principal source of trouble is obvious: it is always necessary to bear in mind, and fatally easy to forget, the number of different questions about punishment which theories of punishment ambitiously seek to answer. I thought when I wrote the first essay in this volume that all that was necessary to dispel the mist from the idea of retribution was to identify these different questions. But I now see that it is necessary also to stress the fact that, at least in the broader modern use of the term 'retribution', there are many different answers to each of these questions, which may be styled 'retributive' and have often earned the title of 'retributive' for the theory of which they form part, even if the theory also contains reformative or deterrent elements normally contrasted with retribution. It is, of course, also true that a stricter or narrower usage of the term still survives, and some writers only allow the title of 'retributive' to theories which give a retributive answer to all the main questions to which a theory of punishment is addressed.

2. A MODEL OF RETRIBUTIVE THEORY

It is I think helpful to start with a simple, indeed a crude, model of a retributive theory which would satisfy this stricter usage. Such a theory will assert three things: first, that a person may be punished if, and only if, he has voluntarily done something morally wrong; secondly, that his punishment must in some way match, or be the equivalent of, the wickedness of his offence; and thirdly, that the justification for punishing men under such conditions is that the return of suffering for moral evil voluntary done, is itself just or morally good. So the theory gives a retributive answer to the three questions, 'What sort of conduct may be punished?', 'How severely?', and 'What is the justification for the punishment?'

Few people would now advocate so thoroughgoing a variety of retribution, or think it reasonable for a legal system to conform to it, especially if we add to it (as Kant did), to avoid the serpent-windings of Utilitarianism, a further feature: that the satisfaction of the conditions required by the theory does not merely make the punishment of the offender permissible, but makes it obligatory, even on the eve of a dissolution of a society against whose laws the person to be punished has offended. But though this model of retributive theory may well be a parody of modern retributivism, it is, I think, illuminating to classify theories which are now termed retributive (either by their advocates or critics) by reference to the ways in which they vary from this severe model.

The range of such theories is very great. I have been astonished to find that Lady Wootton's theories, which I examined in the last two of these essays, are spoken of by some as retributive. This is surprising because Lady Wootton not only criticizes the doctrine of *mens rea* and hopes that it will wither away, but looks forward to the day when the sentence of a criminal court will no longer be thought or spoken of as punishment. To many such a theory, with its great emphasis on the forward-looking aims of penal treatment, and its abandonment of any concern with the mind or will of the offender as it was at the time of the offence, as a condition of liability to conviction, seems the very antithesis of retribution. But is is not quite at the extreme point; for there are those who would wish to eliminate one element that distinguishes the official treatment of crime advocated by Lady Wootton from pure social hygiene, and constitutes a last tenuous connexion between her theory and what would still be called theories of punishment. This element is the requirement that for conviction and subsequent compulsory treatment there must be an offender who has, at least so far as outward conduct is concerned, committed an offence. From the point of view of pure social hygiene it is absurd to wait until crimes have been committed: where there is reliable evidence of anti-social or criminal tendencies, this is enough to justify compulsory measures. Just as Lady Wootton says (wrongly in my view) that the doctrine of *mens rea* has no place within a system of criminal law which aims at the prevention of crime, and attributes loyalty to that doctrine to lingering traces of retributivism, so those who would go further than she does regard as retribution her insistence on a criminal act (i.e. the outward elements of crime) as a necessary condition of conviction. *A fortiori,* the middle way, which I myself have attempted to tread, between a purely forward-looking scheme of social hygiene and theories which treat retribution as a general justifying aim, has itself been regarded as a form of retributive theory. This is because this middle way not only insists on the restriction of punishment to an offender, but also on the general retention of the doctrine of *mens rea*, and allows some place, though a subordinate one, to ideas of equality and proportion in the gradation of the severity of punishment.

It is, however, clear that current controversy about the role and respectability of 'retributive', as opposed to 'utilitarian', theories is not concerned with these weakened versions of the retribution, but with theories which, while allowing certain modifications or modernization of some features of the model, preserve in some form, as being essential to retribution, the principle that the voluntary doing of what is morally wrong itself calls for the punishment of the offender, and the moral gravity of the offence is in itself a proper determinant of the severity of punishment. I shall therefore consider three main modifications of the model, distinguishing the various forms in which it preserves these essential retributive features.

3. MODIFICATION OF THE MODEL

I. Punishment proportionate to the gravity of the offence. To many the most perplexing feature of the model is its requirement that the punishment should in some way 'match' the crime. The simple equivalencies of an eye for an eye or a death for a death seem either repugnant or inapplicable to most offences, and even if a refined version of equivalence in demanding a degree of suffering equivalent to the degree of the offender's wickedness is intelligible, there seems to be no way of determining these degrees. Hence, instead of equivalence between particular punishments and particular crimes, modern retributive theory is concerned with proportionality. But this idea, as Bentham's elaborate treatment of it shows, is susceptible of both a Utilitarian and a Retributive interpretation. In both interpretations it is concerned with the relationships within a system of punishment between penalties for different crimes, and not with the relationship between particular crimes and particular offences. On the retributive interpretation, the relative gravity of punishments is to reflect moral gravity of offences; murder is to be punished more severely than theft; intentional killing more severely than unintentionally causing death through carelessness. It is to such ideas of proportionality that critics of the sentences passed in the Great Train Robbery case,[6] or the decision of the House of Lords in Smith's case, made their appeal. Of course, the conception of the relative moral gravity of different offences is far from simple, and some of its difficulties and the compromises involved in the rough recognition of it as a determinant of the severity of punishment in English courts were explored in the fifth and seventh of these essays. One ambiguity of the idea of the 'gravity' of the offence as a measure of the severity of punishment deserves special notice here since it gives a further inflexion to the idea of retribution. This is the deeply entrenched notion that the measure should not be, or not only be, the subjective wickedness of the offender but the amount of harm done. It is this form of retributive theory that seems to be reflected in the common practice of punishing attempts less severely than the completed crime, or punishing criminal negligence which has a fatal outcome more severely than the same negligence which does not cause death.

II. Retribution as a justifying aim. The retributive principle embodied in the model, that wicked conduct injuring others itself calls for punishment, and calls for it even if its infliction is not necessary in order to prevent repetition of that conduct by the offender or by others, has been attacked on many grounds. To some critics it appears to be a mysterious piece of moral alchemy in which the combination of the two evils of

[6] *R.* v. *Wilson and Others* (1964) 48 Cr. App. Rep. 329.

moral wickedness and suffering are transmuted into good; to others the theory seems to be the abandonment of any serious attempt to provide a moral justification for punishment. Other critics still regard it as a primitive confusion of the principles of punishment with those that should govern the different matter of compensation to be made to the victim of wrong-doing. In its most interesting form modern retributive theory has shifted the emphasis, from the alleged justice or intrinsic goodness of the return of suffering for moral evil done, to the value of the authoritative expression, in the form of punishment, of moral condemnation for the moral wickedness involved in the offence. This theory, expounded in its most convincing form by Bishop Butler in his Sermon on Resentment, is termed by some of its modern advocates a theory of reprobation rather than retribution. But it shares with other modern retributive theories two important points of contrast with Utilitarian theories; for like the model it insists that the conduct to be punished must be a species of voluntary moral wrongdoing, and the severity of punishment must be proportionate to the wickedness of the offence. But this form of theory has also at least two different forms: in one of them the public expression of condemnation of the offender by punishment of his offence may be conceived as something valuable in itself; in the other it is valuable only because it tends to certain valuable results, such as the voluntary reform of the offender, his recognition of his moral error, or the maintenance, reinforcement or 'vindication' of the morality of the society against which the person punished has offended. Plainly the latter version of reprobation trembles on the margin of a Utilitarian theory, in which the good to be achieved through punishment is less narrowly conceived than in Bentham's or in other orthodox forms of Utilitarianism.

III. Combination and compromise with Utilitarian theory. Finally it remains to be observed that most contemporary forms of retributive theory recognize that any theory of punishment purporting to be relevant to a modern system of criminal law must allot an important place to the Utilitarian conception that the institution of criminal punishment is to be justified as a method of preventing harmful crime, even if the mechanism of prevention is fear rather than the reinforcement of moral inhibition. This recognition sometimes takes the form of a rough division of the field as follows. It is insisted that in the considerable and crucially important area of conduct where the prohibitions or requirements of criminal law overlap with morality so that the crime is also a moral offence, it should be a primary concern of the law that punishment should be proportionate to the gravity of the crime, or an adequate expression of moral condemnation for it. On the other hand, it is conceded that there is a vast area of the criminal law where what is forbidden or enjoined by the law is so remote from the familiar requirements of morality that the very word 'crime' seems too emphatic a description of law-breaking. Here the law is what it is, often because of variable and disputable conceptions of social and economic policy; and, in this area, many modern retributivists would concede that punishment was to be justified and measured mainly by Utilitarian considerations. Though most would insist that, even here, the doctrine of *mens rea* should be retained, others might here concede a place for strict liability. This division of the field between retributive and Utilitarian theory is a modern counterpart of the ancient distinctions between *mala in se*, or, as Lord Devlin has put it, 'moral offences with legal definitions attached', and *mala prohibita* which may be regarded as 'quasi crimes'.

In addition to this division of the field other forms of partial accommodation to Utilitarian theory are to be found. The fiercest form of our model of retributive theory was mandatory in the sense that it not merely permitted but required a punishment appropriate to the wickedness of the offence. Some modern retributivists would dissent from this and for them the satisfaction of the conditions constitutes no more than a licence to punish the offender, as one who is morally blameworthy and so punishment-worthy; but whether in this case, he should actually be punished is a question to be settled by reference to the effects which punishment is likely to have on the offender or on the fabric of law and morality in general.

Similar relaxations of the strict requirements of the model may be made in relation to the questions of the amount or severity of punishment, and in the interpretation given to the notion of a proportionate punishment. The sterner forms of retributive theory would regard the moral evil of the offence as justifying a more severe sentence than would be required on deterrent or other Utilitarian grounds: indeed the point is often made that no greater punishment may be needed to deter a murderer than a robber, yet most systems of punishment show their allegiance to retributive ideas by punishing the murderer more severely. But, as was evident in the debates on capital punishment in the House of Lords, many self-styled retributivists treat appropriateness to the crime as setting a *maximum* within which penalties, judged most likely to prevent the repetition of the crime by the offender or others, are to be chosen.

The above does not by any means complete the tale of the variants to be found in current literature or debate on the retributive idea. But it is perhaps enough to serve as a *vade mecum* for the exploration of this now very extensive territory.

NOTES

THESE notes, which vary in length from a few lines to several pages, are designed to bring to the reader's attention criticisms of the views expressed in the text and, in some cases, developments or modification of these views which I now wish to make in the light of criticisms. They also include an account of changes or developments in the law relevant to the matters discussed in the text, and, in the case of the essay on Murder and the Principles of Punishment (Chapter III), a summary of the statistics for the period since the original publication of that essay.

References to the text of this book are indicated by *supra* followed by page numbers, and references to these notes are distinguished by the insertion of the word Notes before page numbers.

The following abbreviations are used:

A.L.I.	American Law Institute
C.L.R.	*Columbia Law Review*
H.C.Deb.	*Hansard:* Parliamentary Debates, House of Commons
Crim.L.R.	*Criminal Law Review*
H.L.Deb.	*Hansard:* Parliamentary Debates, House of Lords
J.C.C.P.S.	*Journal of Criminal Law, Criminology and Police Science*
L.Q.R.	*Law Quarterly Review*
M.L.R.	*Modern Law Review*
P.A.S.	*Proceedings of the Aristotelian Society*

CHAPTER I

Page 3. *Multiple issues and single principles.* Some critics dispute the view urged here that independent principles are relevant at different points in any morally tolerable theory of punishment. See especially M. Goldinger, 'Punishment, Justice and the Separation of Issues', *The Monist*, Vol. 49 (1965), where the alleged separateness of the various issues which I have distinguished is controverted.

Page 4. *Locke's chapter on property.* Though it is important to distinguish between the four questions relating to property which are here distinguished (Definition, General Justifying Aim, Title, and Amount), the late Mr. G. A. Paul convinced me that my criticism here of Locke for failure to distinguish them is mistaken. Mr. Paul contended in his unpublished lectures that Locke (i) provides no definition of property but takes the concept as understood; (ii) uses the notion of 'the labour of a man's body and the work of his hands' only to answer questions concerning the *title* to private property; (iii) gives as the General Justifying Aim of the institution of private property that 'things may be of use to some particular man' and 'do good for the support of his life'; (iv) gives as the criteria of amount or extent of private property 'not spoiling in his possession' and 'enough and as good still left in common'.

The definition of punishment. It has been urged that the definition of punishment given here is defective because it does not include any reference to the fact that punishment is a conventional device for the expression of attitudes of resentment and indignation, and of judgments of disapproval and of reprobation. See J. Feinberg, 'The Expressive Function of Punishment', *The Monist*, Vol. 49 (1965), 1. Bishop Butler in his Sermon on Resentment (1729) treats punishment as the natural expression of 'deliberate resentment' and insists on the importance of the resentment of injustice as one of 'the common bonds by which society is held together'. Cf. also P. F. Strawson, 'Freedom and Resentment', *Proceedings of the British Academy*, Vol. 48 (1962), p. 187.

Page 6. *Distinction between punishment and taxes.* Holmes urged that damages for breach of contract or tort were best treated as taxes on a course of conduct, and at times thought, though with some hesitation, that punishments could also be viewed in this way. See Holmes, *The Common Law*, p. 300, and the *Pollock-Holmes Correspondence*, Vol. 1, pp. 21, 119, 177, and Vol. II, pp. 55, 200–34; M. de W. Howe, *Justice Holmes, The Proving Years*, pp. 74–80. Kelsen, in the latest version of his theory (*Théorie pure du droit* (Paris, 1962), pp. 33–39) attempts to distinguish coercive *sanctions* applicable to delicts (i.e. legal wrongs), from other merely 'administrative' coercive measures, on the purely formal ground that the latter, unlike the former, are not applied to a person whose voluntary act is a condition of their applicability. Cf. Bentham's discussion of the differences between taxation and fines, Bentham, *Principles of Penal Law* (*Works*, Bowring edn., Vol. I, at p. 394).

Page 13. *Justification and excuses.* See, for this distinction, J. L. Austin, 'A Plea for Excuses', 57 *P.A.S.* (1956–7); Bentham, *Limits of Jurisprudence Defined*, 215, n. 37, and 236; Hart, *The Concept of Law*, pp. 174–5; Brandt, *Ethical Theory*, p. 471.

Page 15. *Mitigation and Temptation.* Bentham would not have considered that exposure to an unusual, or specially great, temptation was in itself a ground for mitigating the penalty. See *Principles of Penal Law* (*Works*, Vol. I, 399–400), and *The Principles of Morals and Legislation*, Chap. XIV, paras. 8-9, where he argues that a greater penalty may be required to counter the greater temptation, though on utilitarian grounds the penalty may be mitigated in certain circumstances, depending on the temptation, if these indicate 'benevolence' in the offender, as when a man steals to feed his starving family.

Page 17. *Mental abnormality and diminished responsibility.* s. 2 of the Homicide Act, 1957 (as to which see also Notes p. 246) provides a lesser penalty than that fixed for murder for those whose abnormality of mind has substantially impaired their 'mental responsibility'. This provision has been criticized as incoherent on the grounds that a man must either have been responsible or not responsible at the time he committed a particular crime, and in the latter case should be excused from all penal measures. See Sparks, ' "Diminished Responsibility" in Theory and Practice', 27 *M.L.R.* (1964), p. 9.

Page 19. *Utilitarian justification for excuses.* Efforts have been made to show, contrary to the argument in the text, that the restriction on the use of punishment to those who have voluntarily broken the law is explicable on purely Utilitarian lines. See D. F. Thompson, 'Retribution and the Distribution of Punishment', *Philosophical Quarterly* (1966), p. 59; and T. L. S. Sprigge, 'A Utilitarian Reply to Dr. McCloskey', *Inquiry* (1965), p. 264.

Page 20. *Defences of strict liability on utilitarian principles.* In addition to the arguments in the text, strict liability can in some cases be defended as a means of deterring those who cannot guarantee success in conforming to legal requirements from entering into occupations to which legal strict liability is attached. See R. A. Wasserstrom, 'Strict Liability in the Criminal Law', *Stanford Law Review* (1960), p. 731. It has also been argued that whereas strict liability may stimulate individuals or commercial enterprises to invent new techniques for avoiding violations, this effect cannot be secured merely by penalizing negligence. For the argument that the punishment of negligence is itself a form of strict liability, see Wasserstrom, op. cit. and *contra*, Chap. VI, *supra*.

Nullification and popular conceptions of justice. Bentham thought it most important that punishment should not be 'unpopular' and, though he did not rest his *rationale* for the admission of excuses on the grounds suggested here, was of the opinion that the satisfaction given by punishment of offenders to the public was an element to be taken into account in assessing its utility. See D. F. Thompson, *supra;* Bentham, *Introduction to the Principles of Morals and Legislation,* Chap. XV, paras. 22–24; *Principles of Penal Law,* Part II, Book 1, Chap. X. 'On Popularity' (*Works,* Vol. I, p. 411).

Page 26. *Reform as the general aim of punishment.* I now think that the argument presented in the text that reform *could* not be a General Justifying Aim of the practice of punishment is unsound. Though it would indeed be socially mischievous to subordinate all other considerations to that of reform, such a policy is not logically incoherent. It is *possible* that the actual experience of the pains of punishment may lead to what is usually meant by 'reform', viz. a change of heart and effective resolution to conform to law not because of fear of repeated punishment but out of moral conviction. The main objection in the text, that assigning to reform this place in punishment would subordinate the prevention of first offences to the prevention of recidivism, may be met (partly at least) in different ways. Thus, it has been argued by Lady Wootton (*Crime and the Criminal Law*, p. 101) that since we know so little of the effect on potential offenders of punishment of the guilty, we should normally give priority in the choice of sentence to the likely effect of a particular decision upon the offender. Others have argued that the application of punishment to an actual offender, by marking the law's condemnation of a crime, may not merely deter potential offenders through fear but may strengthen their moral inhibition against the conduct thus condemned, and this, too, may be considered a species of 'reform'. See J. Plamenatz: 'Responsibility', in *Philosophy, Politics and Society*, ed. Laslett and Runciman, 3rd series, Feinberg, op. cit. (Notes, p. 239), and related theories advanced by A. C. Ewing in *The Morality of Punishment*, and Lord Denning, in the *Report of the Royal Commission on Capital Punishment*, Cmd. 8932, para. 53. For criticism of the latter, see *supra*, p. 170-3.

CHAPTER II

Page 28. Criticisms of the main argument in this essay are made by E. L. Beardsley, in Hook, *Determinism and Freedom in the Age of Modern Science*, p. 133; by D. F. Pears, in a review article in *Ratio*, Vol. V (1963), 217–9; and F. R. Berger, in 'Excuses and the Law'. *Theoria* (1965) p. 9.

Page 29. *Determinism and choice*. For the argument that it is logically impossible for a man both to claim to know on inductive grounds what he will try to do on a specific occasion and also to regard the matter as one for his deliberation and choice, see Hampshire, *Thought and Action*, *passim*, and in *Freedom and the Will* ed. Pears (London, 1963), Chap. 6; cf. R. F. Taylor, 'Deliberation and Foreknowledge', in the *American Philosophical Quarterly*, Vol. 1 (1965).

Page 32. *Knowledge and self-control*. English law relating to homicide now recognizes that a man's capacity to conform to the requirements of the law may be substantially impaired, even though he knows at the time of his offence that what he is doing is forbidden by law, see Homicide Act, 1957, s. 2. as interpreted by Parker, C. J., in *R.* v. *Byrne*, [1960], 2 Q.B. 396. Many jurisdictions in the United States and the Commonwealth recognize this more generally by their supplementation of the M'Naghten Rules with clauses providing for a diminished or absent capacity to control actions: see A.L.I. *Draft Penal Code*, Tentative Draft No. 4, pp. 161–9. *Difficulties of proof*. These seem to have been the main reasons that made the House of Lords, in the much criticized decision in *D.P.P.* v. *Smith* (1961), A.C. 290, apply (with modifications) the theory of objective liability expounded by Holmes in *The Common Law* (Lecture 2) to cases of homicide. Holmes's own reasons were, however, quite independent of these difficulties of proof (see Notes, p. 242-4).

Page 34. *Strict Liability*. It is not clear how strict strict liability in Anglo-American law is, e.g. there is a singular lack of authority as to whether it excludes the excuses of duress and insanity. Many theoretical writers insist in general terms that for any form of criminal liability there must be an 'act of will' and that it is on this ground that liability is excluded in cases of 'automatism' and other cases where the accused lacks the minimum control over bodily movements. For an analysis and criticism of this general doctrine see chap. IV, *supra*.

Page 36. In the recent edition (1960) of his *Principles of Criminal Law*, Professor Hall has reaffirmed his view that it is just to punish only 'the intentional doing of a morally wrong act' (pp. 83, 103) proscribed by law. It is, however, not entirely clear to me what the criteria for 'morally wrong' in this context are. Much of Professor Hall's writing suggests that the agent's own moral beliefs are irrelevant, but that it is a necessary condition of just punishment that his actions be wrong according to the accepted or conventional morality of his society. If this is so, it is enough on Professor Hall's view to justify punishment (a) that such moral wrongdoing should have been forbidden by law and (b) that the agent must have intended to do what the law thus forbids.

Pears (op. cit., *Ratio* (1963), p. 219) argues convincingly that though it may be confusing to represent 'moral culpability' as a necessary condition of just punishment—since that suggests that the law should treat something as a crime only if it is already morally wrong—nonetheless, intentional law-breaking does seem to us worse than unintentional law-breaking, and this itself may justify us in recognizing certain excuses quite independently of the reasons presented in the text.

Page 38. *Holmes's theory of objective liability: The Common Law: Lecture 2.* In his famous lecture, Holmes's *idée maîtresse*, which in the end became something of an obsession, was the principle that though the law often seems to make liability to punishment or to pay compensation for harm done dependent on the individual's actual intention to do harm, this is most often not to be taken at its face value. Here, he thought, lay one of the cardinal differences between early and modern law, 'acts should be judged by their tendency under the known circumstances not by the actual intent which accompanies them'; 'Though the law starts from the distinctions and uses the language of morality it necessarily ends in external standards not dependent on the actual consciousness of the individual.' Or again, 'the law considers what would be blameworthy in the average man, the man of ordinary intelligence and prudence, and determines liability by that'. These were, indeed, powerful heuristic maxims dissipating much misunderstanding, especially in the fields of contract and tort. But Holmes came to regard them as more than valuable pointers to neglected tendencies in the law. He sometimes treats them as statements of necessary truths ('by the very necessity of its nature the law is continually transmuting moral standards into external or objective ones'), and he erects these principles into a form of social philosophy justifying what he describes as 'the sacrifice of the individual'.

Such was Holmes's greatly debated theory of objective liability. Its central contention is that when the law speaks of an intention to do harm as a necessary constituent of a crime, all it does, and can, and should require (these three things are never adequately discriminated by Holmes) is that the person accused of the crime should have done what an average man would have foreseen would result in harm. In spite of its subjective and moralizing language, the law does not require proof of the accused's actual wickedness or actual intention or actual foresight that harm would result. Of course, for common sense, as for the law, there are important *connexions* between the proposition that a man in acting in a certain way intended harm, and the proposition that an average man who acted in that way would have foreseen it or intended it. For the latter is good, though not conclusive, evidence for the former. Nonetheless, the two propositions are distinct. Holmes, however, though well aware of the distinction, thought that in general the law did not, and should not attend, to it. This was not because he was a philosophical behaviourist or because he thought that subjective facts were too elusive for the courts to ascertain. There is no echo in Holmes of the medieval Chief Justice Brian of the Common Pleas: 'The thought of man is not triable; the devil alone knoweth the thought of man.' Though many of Holmes's followers accepted his theory of objective liability because of the difficulties of legal proof of actual knowledge or intention, Holmes does not rest his doctrine on these merely pragmatic grounds, but on a social theory. Objective liability for Holmes meant not an evidential test, but a substantive standard of behaviour. His view was that the function of the criminal law was to protect society from harm, and in pursuit of this objective it did, and should, set up 'objective' standards of behaviour, which individuals must attain at their peril. The law may exempt those who, like the young child or lunatic, are obviously grossly incapable, but apart from this, if men are too weak in understanding or in will-power, they must be sacrificed to the common good.

Certainly the criminal law bears traces of such objective standards; indeed, the elimination of these has been the aim of many liberal minded reformers of the law for many years. But though Holmes at one point says that he does not need to defend the law's use of 'objective standards' but only to record it as a fact, he devotes much of this lecture to showing that the law is reasonable and even admirable. The arguments he uses are the poorest in his book. He considers the objection that the use of external standards of criminal responsibility, taking no account of the incapacities of individuals, is to treat men as things, not as persons, as means and not as ends. He admits the charge but thinks it irrelevant. He asserts that society frequently treats men as means: it does so when it sends conscripts 'with bayonets in their rear' to death. But this reply is cogent only against a stupidly inaccurate version of the Kantian position on which the objection rests. Kant never made the mistake of saying that we must never treat men as means. He insisted that we should never treat them *only* as means 'but in every case as ends also'. This meant that we are justified in requiring sacrifices from some men for the good of others only in a social system which also recognizes their rights and their interests. In the case of punishment, the right in question is the right of men to be left free and not punished for the good of others, unless they have broken the law when they had the capacity and a fair opportunity to conform to its requirements.

Apart from this, Holmes's main argument is a fallacy and, unfortunately, an infectious one. He adopts the acceptable position that the general aim justifying a modern system of criminal punishment is not to secure vengeance or retribution in the sense of a return of pain for an evil done, but is to prevent harmful crime. On this basis he seeks to prove that there can be no reason why the law should concern itself with the actual state of the offender's mind or enquire into his actual capacity to do what the law requires. His proof is, that since the law only requires outward conformity to its prescriptions and does not care, so long as the law is obeyed, what were the intentions or motives of those who obeyed, or whether they could have done otherwise, so it should equally disregard these subjective matters in dealing with the offender when the law has been broken. This is of course a *non sequitur*. Even if the general justification of punishment is the utilitarian aim of preventing harm, and not vengeance or retribution, it is still perfectly intelligible that we should defer to principles of justice or fairness to individuals, and not punish those who lack the capacity or fair opportunity to obey. It is simply not true that such a concern with the individual only makes sense within a system of retribution or vengeance. Holmes himself, indeed, in discussing liability in tort, stresses the importance of such principles of justice to individuals, but thinks that in the criminal law their requirements are adequately satisfied if the individual is punished only for what would be blameworthy in the average man. No doubt there are practical difficulties in ascertaining the actual knowledge or intention or capacity of individuals in every case, but there is no reason in principle why a maximum effort should not be made to do it.

Page 39. *Questions for legislatures and questions for judges.* This way of presenting the distinction between the question whether the laws are good enough to justify their enforcement, and the question whether they are justly applied in particular cases, may be misleading because judges, or juries under the instruction of judges, have not only had power to determine the fact of violations of the criminal laws and the sentence, but also power to create new offences (see Glanville Williams, *Criminal Law, The General Part,* 2nd edn., Chap. XII, pp. 592–608, and the revival by the House of Lords, in *R.* v. *Shaw* (1962), A.C. 220, of the common law offence of a conspiracy to corrupt public morals). In any case the indeterminacy at the border lines of legal rules blurs the distinction between legislative and judicial questions.

Page 47. *The law as the guide of individuals' choices.* It should be observed that the law's threat of punishment need not operate only in the way suggested in the text, i.e. by entering into the deliberations of a person tempted to commit a crime. It is also the case that the threat of punishment, or the knowledge that other persons have been punished, may stimulate a man to a greater exercise of his faculties, wakefulness and care, so that he does not commit an offence through negligence or inadvertence. Some scepticism of the possibility of the law operating to prevent negligent conduct is due to exclusive concentration on the place that threats have in the conscious deliberations of a person contemplating committing a crime. For further discussions of these issues see Chapter VI.

Page 49. *Excusing conditions as protection of the individual against the claims of society for the highest measure of protection from crime.* This concern for the individual may be represented, along the lines familiar to economists, of maximizing a variable subject to restraints, as a restraint upon the maximizing of the value constituted by the prevention of harmful crime. (See Barry, *Political Argument*, pp. 4–8 esp. p. 5, n. 2); the admission of strict liability in some cases will represent a decision to prefer an extra measure of security to justice to the individual (the restraint), whereas its exclusion represents a decision to prefer justice to the individual to the higher measure of security.

Page 52. *Punishment and social hygiene.* See, for the most recent advocate of the substitution of what is, in effect, a system of social hygiene for punishment, Barbara Wootton, *Crime and the Criminal Law*, Chap. II *et. seq.*, criticized in Chapters VII and VIII.

CHAPTER III

Page 54. *Murder and its punishment in England, 1957–64.* This and the next note describe the principal developments in England during this period: for fuller details and an illuminating statistical analysis of the period 1957–60 see, *Murder*, Home Office Research Unit Report, by E. Gibson and S. Klein (H.M.S.O., 1961). For the period since 1964 see Appendix, p. 268 *infra*.

(*a*) *The operation of the Homicide Act, 1957.* The two main changes introduced by this Act were the restriction of the death penalty to capital murders, as defined by s. 5 of the Act (specified *supra*, p. 57, n. 8), and the introduction of the defence of diminished responsibility as a ground for reducing murder to manslaughter in those cases where the accused was 'suffering from such abnormality of mind as impaired his mental responsibility for his acts or omissions in doing the killing' (s. 2). These changes were followed in the years 1957–64 by a rise in the recorded rates of murder from an average of 3.1 per million of population for the years 1951–5 to an average of 3.8 (i.e. an increase of 26 per cent (See 716 *H.C.Deb.* 416 (1964–5)). But even if the cases reduced to manslaughter on the grounds of diminished responsibility under s. 2 are counted, as they are in these figures, as murder, the rate at its worst (1963-4) never exceeded 4 per million of population. It seems very improbable that any significant increase in the rate of murder can be attributed to the Homicide Act, since the proportion of all murders classed as capital and carrying the death penalty after the Act (13.5 per cent for the years 1957–64) was very similar to the estimated proportion of murders (14.4 per cent) which in the five years 1952–7 would have been regarded as capital had the distinction between capital and non-capital murder been then in operation (see Gibson and Klein, op. cit., p. 8, and the discussion in the House of Lords on this point in 1965, 268 *H.L.Deb.* 492, 504, 702 (1964–5) and 269 *H.L.Deb.* 545, 552 (1964–5).

More surprisingly the revision of s. 2. of the Act, supplementing the defence of insanity as defined by the M'Naghten Rules with the new plea of diminished responsibility, was not followed by any increase in the proportions of persons escaping convictions for murder on the ground of mental abnormality. It appears that to a large extent the plea of diminished responsibility replaced the old plea of insanity and the combined totals of those convicted and found insane or suffering from diminished responsibility after the Act constituted practically the same proportion of those convicted as did the total of those convicted and found insane under the M'Naghten Rules for the years 1952–6, before the Act (see Gibson and Klein, op. cit., pp. 8–10). The courts in the first years after the introduction of the defence of diminished responsibility were disposed to construe it very narrowly, and refused to indicate to juries that inability to control behaviour as distinct from mere lack of knowledge came within the scope of the new defence (see *R.* v. *Spriggs,* (1958), 1 Q.B. 270). But in *R.* v. *Byrne,* (1960), 2 Q.B. 396), the Lord Chief Justice emphasized that under the defence of diminished responsibility 'not only the perception of physical acts and matters and the ability to form a rational judgement whether an act is right or wrong, but also the ability to exercise will-power to control physical acts in accordance with that rational judgement', were matters to be considered.

The Homicide Act, though it had little effect on the rates of murder, drastically reduced the number of executions to an average figure of 4 per annum for the period 1957–64 (only 2 in each of the three years 1962-4, as compared with an average of about 13 in the years 1952-5 (see 268 *H.L.Deb.* 463 (1964–5)) and Gibson and Klein, op. cit., p. 10, Table 7.)

The principal statistics for the period 1957–64 relating to the issues discussed in the text may be summarized as follows:

Year	*Nos. of murders known to police* (incl. s. 2 cases)	*Rate per million of population*	*No. of executions*
1957	157 (22)	3.5	3
1958	143 (29)	3.2	5
1959	156 (21)	3.4	4
1960	154 (31)	3.4	7
1961	148 (30)	3.2	4
1962	171 (42)	3.7	2
1963	178 (56)	3.8	2
1964	170 (35)	3.6	2

The figures in brackets represent the numbers of s. 2 cases included in the figures to which they are annexed. All the above figures are taken from Gibson and Klein, *Murder 1957 to 1968* (H.M.S.O., 1969), p. 2, table 1, and p. 11, table 8. N.B. ibid., p. 1, para. 8, and p. 62, for explanation of the adjustment of figures given in Gibson and Klein, *Murder* (H.M.S.O., 1961).

(*b*) *The Murder (Abolition of Death Penalty) Act, 1965* If the figures for the years 1957–64 did little to confirm fears that the restriction of the death penalty to the capital murders specified in s. 5 of the Homicide Act would lead to a significant increase in the murder rate, nonetheless, the distinction which the Act drew between capital and non-capital murders was widely criticized as an anomaly and injustice. Dissatisfaction with the operation of the Act in this respect eventually led to a consensus in favour of total abolition, quite without precedent in England, between a hitherto mainly conservative judicial opinion and the opinion of those who had long fought for abolition. Thus Lady Wootton, an abolitionist of long standing, described the compromise between abolition and retention of the death penalty, represented by the Homicide Act as 'a disastrous failure', on the ground that it had produced great anomalies and had also failed to do what its authors intended, viz., to secure that 'the professional criminal would be the person whose crimes would attract the death penalty' (268 *H.L.Deb.* 459 (1964–5)), and the Lord Chief Justice made his own conversion to abolition plain in the following words. 'I am in favour of abolition, not I am afraid, on any moral ground, but merely because of the working of the Homicide Act, 1957. I confess, looking back eleven years, that if anybody had said that I should come out as a full-blooded abolitionist I should have been surprised.' He added that he and 'all the judges' were 'quite disgusted' with the results produced by the Homicide Act, such as that the taking of a coin or a note made all the difference between capital murder and non-capital murder, and persons equally blameworthy were frequently differently treated under it. (See 268 *H.L.Deb.* 480–1, (1964–5)).

These opinions were expressed in the debate in the House of Lords in July 1965 on a bill designed to substitute imprisonment for life for the death penalty in all cases of murder. This bill had been introduced in the House of Commons in 1964, and the government, though it did not adopt the bill, had announced that it would provide facilities for 'a free decision of Parliament' on the issue. The bill became law as the Murder (Abolition of Death Penalty) Act, 1965, on 9 Nov. 1965, after many Parliamentary hazards and delays. Two important amendments were made to it in its passage through Parliament. The first of these (s. 4) was designed to secure for the Act an experimental character, and provides that the Act shall expire on 31 July 1970 unless both Houses of Parliament resolve otherwise. The second amendment (ss. 1(2) and 2) was designed to restrict the Home Secretary's powers to determine the length of imprisonment to be served by those convicted of murder and sentenced to imprisonment for life. These sections provide (s. 1(2)) that the court in sentencing a person convicted of murder may declare the period which it recommends to the Home Secretary as a minimum period to elapse before release on licence, and (s. 2) that the Home Secretary shall consult the Lord Chief Justice together with the trial judge, if available, before releasing a person convicted of murder. It is to be noted that there is nothing in these provisions to compel the Home Secretary to follow the recommendations or advice of the judges. During the debates strenuous but unsuccessful efforts were made to give the courts power to fix, as in manslaughter, a determinate sentence to be served, up to the maximum sentence of imprisonment for life, instead of forcing them to pass sentence of imprisonment for life, which is in effect an indeterminate sentence at the discretion of the Home Secretary. These efforts were the outcome of anxiety lest the comparatively short periods of imprisonment (under ten years) served by the majority of those murderers whose sentence had been commuted to imprisonment for life when the death penalty was operating might be taken by potential murderers to be the standard maximum. For the continuance of the Murder Act, 1965, see Appendix, p. 268 *infra*.

Page 54. *Murder in the United States, 1957–64*. For a general survey, see H. A. Bedau, *The Death Penalty in America* (1964); A.L.I. *Model Penal Code*, tentative draft No. 9 (1959), esp. the *Report on the Death Penalty* by Thorsten Sellin, appended to this volume; and Patrick, 'The Status of Capital Punishment: A World Perspective', *J.C.C.P.S.* (1965) p. 397. The relevant statistics for the years 1957–64 are as follows:

	Murder and non-negligent manslaughter	*Rate per million of population*
1957	8,060	47
1958	8,220	47
1959	8,580	48
1960	9,000	50
1961	8,630	47
1962	8,430	45
1963	8,530	45
1964	9,250	48

(The figures up to 1959 are taken from the summaries in *Uniform Crime Reports*, 1957–64, as to which, see Notes, p. 250. N.B. *Uniform Crime Report*

(1959), p. 4 for adjustment of earlier figures. The figures for 1960–4 are taken from the summary in the *Uniform Crime Report*, 1968. For later figures see Appendix, p. 268 *infra*.)

Legislative changes in relation to the death penalty. There is at present (1966) no death penalty in Maine, Michigan, Wisconsin, Minnesota, Alaska, and Hawaii, and, except for murder committed by those under sentence of imprisonment for life, in North Dakota and Rhode Island. The death penalty was abolished in Delaware in 1958 but was restored in 1965 after two murders committed by negroes (see Bedau, op. cit., p. 359). The new states admitted to the Union (Hawaii and Alaska) had abolished the death penalty in 1957 before admission. It was abolished in Puerto Rico in 1929 and in the Virgin Islands in 1957.

Page 56. *English public opinion and the death penalty.* In both Houses of Parliament during the debates in 1965 opponents of the bill to abolish the death penalty objected on the ground that popular opinion was firmly against abolition. It was argued that since the three main national opinion polls held in 1965 showed that 65 per cent were against abolition and only 20 per cent in favour, and since no Member of Parliament had mentioned the death penalty in his election address, the bill was 'not a democratic reflection of the wishes of our people' (Lord Colyton, 269 *H.L.Deb.* 531–3 (1964–5) cf. E. Gardiner (716 *H.C.Deb.* 438 (1964–5)). Against this interpretation of parliamentary democracy, there were, in both Houses, reassertions of the principle that members of parliament were not delegates but representatives and of the duty of the House of Lords to lead public opinion. Lady Wootton repudiated 'the unusual argument that [parliamentary decisions] should be governed by the Gallup Poll' (268 *H.L.Deb.* 462-3 (1964–5)), and the Archbishop of York asked 'if we are not here to give a lead, what are we here for?' (269 *H.L.Deb.* 538 (1964–5)). In the General Election of 1966 an Independent candidate standing solely on the death penalty issue, opposed Mr. Sydney Silverman, the principal protagonist of abolition during the last twenty years, and polled 4,577 votes—well over one eighth of the poll. The candidate was Mr. P. Downey, in the Nelson and Colne constituency; he was an uncle of a child victim of a particularly revolting murder (the 'Moors' murder: *R.* v. *Brady*, *The Times*, 20 April 1966).

Page 59. *Abolition of the death penalty in the U.S.A.* See *supra*.

Page 59. *Mandatory death penalty in the U.S.A. for first degree murder.* Though the death penalty is still mandatory for certain types of murder in some states (e.g. murder by persons serving sentence of life imprisonment), and also for certain other offences, it is no longer so for all first-degree murder. New York, the last state to retain the mandatory death penalty for it, substituted jury discretion in 1961.

Page 62. *Objective standards and subjective tests in murder.* The much criticized decision of the House of Lords in *D.P.P.* v. *Smith* (1961), A.C. 290, applying with certain modifications the doctrine of Holmes of objective liability (see *supra*, p. 38 and Notes pp. 242–4) was regarded by its opponents as a departure from enlightened principles of criminal responsibility and a virtual restoration of the doctrine of constructive murder: see Glanville Williams, 23 *M.L.R.* (1960), p. 606, and *Howard Journal*, (1961), pp. 307-9; for contrary interpretations see Denning, *Responsibility before the Law* (Jerusalem 1961); and for the subsequent history of the Smith doctrine, see Buxton, 'The Retreat from Smith', *Crim.L.R.* (1966), p. 195. For its abrogation by s. 8 of the Criminal Justice Act, 1967, see *R.* v. *Wallett* (1968), 2 Q.B. 267.

Page 63. *Length of imprisonment in murder cases.* Since 1957 much longer sentences have been countenanced in England and periods formerly served by reprieved murderers cannot be regarded as a guide to the periods to be served by those now sentenced to life imprisonment. See, for long sentences for offences other than murder, *R.* v. *Blake* (1961), 45 Cr.App.R. 292 (42 years) and *R.* v. *Wilson and others* (1964), 48 Cr.App.R. 329 (30 years).

Page 64. *Second murder by a released murderer.* See the case of *R.* v. *Simcox* (*The Observer*, 15 March 1964). Simcox, who had been sentenced to death for the murder of his wife, was reprieved and released after imprisonment, and later murdered a second woman. He was sentenced to death for this second murder as a capital murder under the Homicide Act, 1957, but again reprieved. Shortly before the second murder he was convicted of the unlawful possession of firearms but was merely put on probation (see 716 *Comm.Deb.* 384 (1964–5)).

Page 64. Double jeopardy. The Criminal Appeal Act, 1968, s. 2, provides that the Court of Appeal shall allow an appeal if they think the verdict should be set aside on the ground that 'under all the circumstances of the case it is unsafe or unsatisfactory' or 'that on any ground there was a material irregularity in the course of the trial', subject to the proviso that they may dismiss the appeal if they consider that 'no miscarriage of justice' has actually occurred. (The word 'substantial' appeared in the formulation of the similar proviso in the Criminal Appeal Act, 1907, s. 4.)

Page 66. *The Uniform Crime Reports.* For criticisms of the criminal statistics in these reports see Beattie, 'Criminal Statistics in the United States', *J.C.C.P.S.* (1960), p. 49. Bedau, op. cit., pp. 56–120. Lejins, 'The Uniform Crime Reports', *Michigan Law Review* (1966), p. 1011, and Robison, 'A Critical View of the Uniform Crime Reports', ibid., p. 1031. The figures given in the Uniform Crime Reports cover 98 per cent of the U.S. population in 'metropolitan' areas, but only 94 per cent of the total national population, not 98 per cent of the total, as stated in Bedau, op. cit., p. 63 (see *Uniform Crime Report* (1962), p. 28).

Pages 68–9. *Comparison between English and American rates.* (See tables *supra* for the national totals and rates.) For a comparison between areas similar to those described on p. 69, the figures for 1962–4 were as follows:

(i) *Rates of State total in Georgia*: For the years 1962–4: 103, 94, 117, murder and non-negligent manslaughter per million of population; to be compared with the rates in New Hampshire for the same years, viz. 24, 32, 9 per million of population.

(ii) *In Chicago:* For the years 1962–4, the *number* of murders and non-negligent manslaughters were 442, 424, and 468, representing rates per million of population of 70, 67, and 72, respectively: the English figures for the same years for murders (including diminished responsibility cases) were: 179, 189, 190, representing rates per million of population of 3.8, 4.0, and 4.0.

Page 70. *Numbers of executions.* In the five-year period 1958–62, the average number of executions in the United States (for all capital crimes, not only murder) was 48.6, compared with 82.6 for the years 1950–4, and 127.8 for the years 1945–9. (See 'Executions', *National Prisoner Statistics,* No. 32 (April 1962), cited in Patrick, op. cit., at p. 409).

For the five-year period 1958–62 the average number of executions in the U.K., according to the *Criminal Statistics: England and Wales,* was 4.4, and not 5.0 as stated in Patrick, op. cit., at p. 400.

Page 70. *Murder by professional criminals.* For detailed discussion of the various motivations for murder see Wolfgang, *Patterns in Criminal Homicide* (1958), and *A Sociological Analysis of Criminal Homicide* (1961), in Bedau, op. cit., p. 74. These and other studies of the figures now available suggest that criminal homicide in the United States and in England usually results from quarrels, jealousy, arguments over money, and robbery. In Philadelphia for the period 1958–61, close friends or relatives were victims in over half the cases. This was confirmed by later investigations in Philadelphia of all criminal homicides for the years 1958–61 (see Pokorny, 'A Comparison of Homicides in Two Cities', 56 *J.C.C.P.S.* (1965), p. 479). Two-thirds of the offenders in the earlier investigations had a previous arrest record.

In England, in the years 1957–60, quarrels and jealousy accounted for more than half the total of those convicted for murder, and 20.7 per cent of those convicted committed murder for financial gain (see Gibson and Klein, op. cit., pp. 32–34).

Page 77. *Strict liability and fairness.* For other objections to strict liability not based on the unfairness of punishing those who were unable to avoid doing what the law forbids, see *supra*, pp. 23-4, 46-9.

Page 81. *Crime and disease.* Cf. Barbara Wootton, *Crime and the Criminal Law* (1963), criticized in Chapters VII and VIII. Lady Wootton's arguments, though they are independent of any philosophical doctrine of determinism, lead to the conclusion that the doctrine of *mens rea* is illogical if the aim of the criminal law is the prevention of socially damaging actions, and not retribution for past wickedness.

Page 83. *Statistical evidence of the superior efficacy of the death penalty as a deterrent.* For a general survey of the evidence see Bedau, op. cit., pp. 258

et seq. The arguments in the text that the statistics justify only the statement that there is no evidence that the death penalty is a superior deterrent (as distinguished from the statement that there is evidence that the death penalty is not a superior deterrent) have appeared to some critics to err on the side of caution. It is argued, against this cautious view, that it confuses what can be truthfully said of any single country with what can be reasonably said of a number of different countries taken together. It is true that in any given case where after abolition of the death penalty the murder rate either failed to rise or fall, the efficacy of the death penalty may be masked in the statistics in the manner suggested by the hypothetical example of England from 1910–30, cited in the text (*supra*, p. 84) from Gold's article. For in any given case where after abolition the rate either failed to rise or fell, it is possible that during the period following abolition some independent cause tending to produce a fall in the rate of murder was operative, and the rate would have fallen further than it actually did had the death penalty not been abolished. But critics of the more cautious view urge that though this is not impossible or improbable in any given case, it is improbable that in *every* case or *very many cases* some such causes, tending to lower the murder rate, should have been operative during the period after abolition so as to mask the effect of abolition. This attractive argument, however, fails to take account of the fact that in very many of the cases where the death penalty has been abolished the rate of murder was already declining before abolition, and in such cases some causes tending to lower the rate must have been already operative; so there is nothing improbable in supposing that such causes were also operative after abolition in all such countries. Had the rate of murder not been declining before abolition, and had it remained stable or fallen after abolition, this critical argument would have had force; for it would be exceedingly improbable that in each abolition country abolition coincided with some *new* factor tending to lower the rate. See Bedau, op. cit., pp. 256 for discussion of this point.

Page 89. *The death penalty and the possibility of error.* In 1948, in opposing the abolition of the death penalty, Lord Kilmuir, then Sir David Maxwell Fyfe, former Attorney-General, said, 'There is no practical possibility' of an innocent man being hanged in this country and that anyone who thinks otherwise is 'moving in a realm of fantasy'. See 449, *H.C.Deb.* 1077 (1947–8). It is, however, now generally agreed that Timothy Evans, who was convicted and executed for the murder of his child in 1950, would not have been convicted had evidence which came to light in 1953 been available. Evans's case has been the subject of numerous parliamentary debates in England besides those cited in the text: see especially 518 *H.C.Deb.* 1435–54 (1953); 642 *H.C.Deb.* 649–711 (1961); 705 *H.C.Deb.* 1256–7 (1965). Evans's case has also been the subject matter of two extrajudicial enquiries: the first of these (conducted by Mr. Scott Henderson, Q.C., in 1953 in the interval between the conviction and execution of John Christie for the murder of a number of women) found, 'for reasons of overwhelming cogency', that there was no good ground to suspect a miscarriage of justice in Evans's conviction. The report of the second enquiry (by Brabin J., Cmd. 3101, October 1966), set up in 1965 as a result of grave dissatisfaction with the first enquiry, found that it was 'more probable than not' that he did not kill his child, for whose murder he was executed but it was 'more probable than not' that he killed his wife. See for further discussion of the doubts concerning Evans's guilt, L. Kennedy, *10 Rillington Place,* to which the Scott Henderson report is appended. But see also for doubts concerning the guilt of a boy of 14 sentenced to death in Canada in 1959, but reprieved though still in prison, I. Lebourdais, *Stephen Truscott* (London, 1966).

CHAPTER IV

Page 90. *Developments since* 1960: Since the publication of this essay the courts have clarified the relationship between insanity as defined by the M'Naghten Rules and 'automatism', which is now the expression generally used to denote the fact that the movements of the body or limbs of a person are involuntary (See *Bratty* v. *A.G. for Northern Ireland* (1963), A.C. at p. 386, and the comments thereon by Cross in *Reflections on Bratty's Case,* 78 *L.Q.R.* (1962), p. 236, and in Cross and Jones, *Introduction to the Criminal Law,* 5th edn., p. 67). The upshot has been the establishment of a clumsy and complex distinction between 'insane automatism', where the accused's lack of control of his bodily movements is due to a disease of the mind, and 'sane automatism', where it arises from some other forms, e.g. concussion, or a sudden illness not amounting to disease of the mind. Insane automatism is governed by the M'Naghten Rules, and the legal burden of proof rests on the accused who, if he is successful on this issue, is ordered to be detained in hospital at the Queen's pleasure; sane automatism, if established, entitles the accused to an acquittal, and the burden of negativing it rests on the prosecution; though the accused must, in order to raise the issue, adduce sufficient *prima facie* evidence to suggest a reasonable doubt that his bodily movements were voluntary, and the courts have emphasized that for this purpose the accused must usually adduce evidence of the cause of the alleged condition (see *Hill* v. *Baxter,* (1958), 1 Q.B. at p. 285, and Bratty's case (1963), 3 A.C. at p. 414). These developments have in part been inspired by the reluctance on the part of the courts to grant an unqualified acquittal to a person who while unconscious has gravely injured another (see for such cases *R.* v. *Charlson* (1955), 1 All E.R. 859, disapproved of by Lord Denning in Bratty's case, and *R.* v. *Boshears, The Times,* Feb. 18, 20, 1961). Similar concern accounts also for the widening of the notion of disease of the mind so as to bring cases of automatism within the M'Naghten Rules (see *R.* v. *Kemp,* (1957), 1 Q.B. 399, at p. 408). For the stringent standards of proof required in cases of sane automatism, see *Watmore* v. *Jenkins,* (1962), 2 Q.B. 572. For parallel developments in Australia, and illuminating comment thereon, see Morris and Howard, 'Insanity and Automatism' in *Studies in Criminal Law* (Oxford, 1964), p. 37, and Howard, *Australian Criminal Law* (1965), pp. 279–83. In Scots law the onus is always upon the accused to establish automatism on the balance of probabilities. See *H.M. Advocate* v. *Cunningham* (1963), S.L.T. 345, overruling on this point *H.M. Advocate* v. *Ritchie* (1926), J.C. 45; cf. *Stevenson* v. *Beatson, Crim.L.R.* (1966), p. 339.

Page 90. *The Courts and the general doctrine,* The general doctrine has now been judicially formulated by Lord Denning: 'in *Woolmington* v. *D.P.P.* Lord Sankey said that the Crown must prove (a) that death was the result of a voluntary act of the accused and (*b*) the malice of the accused. The requirement that there should be a voluntary act is essential not only in a murder case but also in every criminal case. No act is punishable if it is done involuntarily; and an involuntary act in this context—some people nowadays prefer to speak of it as automatism—means an act done by the muscles without any control by the mind as a spasm, a reflex action, or a convulsion, or an act done by a person who is not conscious of what he is doing, such as an act done while suffering from concussion or while sleep-walking' (Bratty's case (1963) A.C. at p. 409.) A general formulation is also to be found in codes in the Commonwealth. The Queensland Code s. 23 provides (with exceptions) that no one is criminally responsible for an act or omission which occurs independently of the exercise of his will (see also the Tasmanian Criminal Code, s. 13 (i) and the Israeli Criminal Code Ordinance, 1936, s. 11, and the wide interpretation given to this requirement in *Mandelbrot* v. *A.G.*, (1965), 10 P.D. 281, so as to include forms of insanity outside the scope of the M'Naghten Rules).

Page 90. The suggestion that the doctrine requiring a voluntary act is only important in cases of strict liability, made here and on p. 107 *supra*, now needs correction. The defence of automatism has been considered as a distinct defence in several cases where the accused was charged with offences in which *mens rea* is a necessary ingredient. (See Bratty's case and *R.* v. *Charlson*, *supra*, p. 253). The explanation of the relevance of automatism in such cases is that though it is true that a person in a state of unconscious automatism must necessarily lack the knowledge, foresight, or intention which constitues *mens rea*, automatism is now regarded as a distinct phenomenon requiring specific investigation, and cannot be treated as ordinary accidents or mistakes which exclude *mens rea*. It is governed by distinct principles as to the burden and standards of proof mentioned above.

Page 92. *The general doctrine and strict liability.* It now seems clear that though the offences of dangerous driving (s. 2 Road Traffic Act, 1960), and failing to give precedence at a zebra crossing (s. 46 of the Road Traffic Act. 1960), are said to be offences of strict or absolute liability, the accused is entitled to acquittal if through no fault of his own he was deprived of control of the car at the relevant time (see *R.* v. *Spurge* (1961), 2 Q.B. 205, and *Burns* v. *Bidder, The Times,* 5 May 1966). In Spurge's case

Salmon L. J. said, (p. 211) 'there does not seem to be any real distinction between a man being suddenly deprived of all control of a motor-car by some sudden affliction of his person, and being so deprived by some defect suddenly manifesting itself in the motor car. In both cases the motor-car is suddenly out of control through no fault of his'. This method of approach seems preferable to treating liability, as Goddard C. J. did in *Hill* v. *Baxter* (loc. cit.), as depending on the question whether the accused in such conditions of automatism could be said to be 'driving' at the material time. It is, however, plain that though the general principle suggested by Salmon L. J. covers both types of cases, exemption from liability where a mechanical defect has put the car out of control cannot be regarded as an application of the general doctrine that a voluntary movement is required for liability. Salmon L. J. based the exemption in such cases on the ground that the driving was only the occasion, and not the cause, of the danger.

Page 101. *Contracting the muscles as an action.* For a consideration of cases where the context and intentions of the agent make it reasonable to describe his action as contracting his muscles, see D'Arcy, *Human Acts*, pp. 4–12.

Page 104. *Reconstruction of the general doctrine.* The suggestions made in the text have been criticized in reviews of a shorter version of this essay published in *Freedom and the Will*, ed. Pears (1963), Chap. 3: see *Mind* (1964), p. 303, *Journal of Philosophy* (1964), p. 308 and pp. 310–12 and especially Houlgate, 'Acts Owing to Ignorance', *Analysis* (October 1966) and Jager, 'Describing Acts Owing to Ignorance', *Analysis* (April 1967).

I do not now consider satisfactory the criterion offered in the text for the identification of the fundamental defect in all those cases, conscious and unconscious, where the law holds the requirements of a minimum mental element not to be satisfied, even where liability is 'strict'. Though this criterion may be generally on the right lines in insisting that the defects must be characterized by reference to the notion of an action which the agent believes himself to be doing, it suffers from the following two faults: (i) the obscurity of the expressions 'not required for an action', 'not appropriate to an action', and 'not forming part of an action'; (ii) doubts as to whether the expression 'an action which the agent believed himself to be doing' might not include an involuntary action of which a man might be fully conscious, such as falling downstairs, or the case where the involuntary tremors of a palsied man break a glass, in which case the suggested criterion would plainly be useless. It may, however, be urged, in reply to the last criticism, that in ordinary usage the expression 'he believed that he was doing a certain action' really does exclude such cases. However, the first criticisms have to be met, and to meet both criticisms I would now substitute for the expressions 'not required for', 'not appropriate' and 'not forming part of', a reference to the agent's *reasons.* The fundamental defective cases would then be those where the bodily movements occurred though the agent had no reason for moving his body in that way, and for this purpose the agent's moving some part of his body 'just because he wanted to do so' would count as a reason for moving it, though in most cases a man's reason would be the performance of some action other than merely moving part of his body.

I appreciate that, against this amended version of the suggestions in the text, it may be said that it assumes, and does not explain, the distinction between a bodily movement occurring and a man making a movement of his body for some reason. It is, however, to be observed that the criterion is not offered as a philosophical analysis of this distinction, but as a means of identifying (without importing any fictional elements) and specifying in a terminology which is in ordinary use the common element in those cases where the law exempts a man from responsibility on the grounds that the minimum mental element required is not present.

It may also be said that this amended version is vulnerable itself to the criticism urged against Austin in the text, viz.: that it assumes unrealistically that the agent is normally aware of the muscular contractions involved in action and desires them before acting. But 'bodily movement' is not the same as 'muscular contraction', and the normal agent is surely aware of what part of his body he moved when he did some action such as kicking a football, and could give his reason for moving his body in that way, without claiming that he had or felt a desire for the bodily movement before acting.

CHAPTER V

Page 114. *Mens Rea.* For divergent views of the proper scope of this term see *supra*, p. 139-40.

Page 115. *Provocation.* If established, provocation in English law reduces a charge of murder to manslaughter, but is not a defence in English law in other crimes (see *R.* v. *Cunningham* (1959), Q.B. 288). In some jurisdictions in the United States certain forms of provocation, if established, entitle the accused to an acquittal (see Texas Penal Code, Art. 1220, which makes the killing by a husband of a wife's paramour taken in the act of adultery 'justifiable homicide'). In some Commonwealth jurisdictions provocation may have a wider scope: see Howard, *Australian Criminal Law*, pp. 82, 129. See also Art. 210. 3(1)*b* of the A.L.I. Model Penal Code.

Duress. The scope of duress as a defence in English law is uncertain and is rarely established (see the remarks of Goddard C. J. in *R.* v. *Steane*, (1947), K.B. 997, at p. 1005, and Edwards, 'Compulsion, Coercion and Criminal Responsibility', 14 *M.L.R.* (1951), p. 297.

Page 115. *Constructive murder and objective tests.* According to the doctrine of constructive murder (felony murder), any killing in the course of the execution of a felony (or in later phases of English law in the execution of a violent felony) was murder. The doctrine was abolished in England by the Homicide Act, 1957, s. 1(2), but is still law in the United States in many jurisdictions, where it is often widely interpreted. (See Morris, 'The Felon's Responsibility for the Lethal Acts of Others', 105 *Univ. of Pennsylvania Law Rev.* (1956), p. 50).

Objective tests: see the decision of *D.P.P.* v. *Smith,* (1961), A.C. 290, and the discussion of Holmes's doctrine of objective liability, *supra*, p. 38 and Notes, p. 242-4.

Malicious damage: the lay reader should note that 'in any statutory definition of a crime, 'malice' must be taken not in the old vague sense of wickedness in general, but as requiring either (i) An actual intention to do the particular kind of harm that in fact was done; or (ii) Recklessness as to whether such harm should occur or not . . . It is neither limited to, nor does it indeed require, any ill-will towards the person injured'. (See *R.* v. *Cunningham* (1957) 2 Q.B. 396 approving this statement by Kenny in *Outlines of Criminal Law,* 16th edn., p. 186). But see *R.* v. *Mowatt* (1967), 1 Q.B. 241.

Page 116. *Intention.* For recent philosophical discussion of intention see G. E. M. Anscombe, *Intention* (Oxford, 1957); and for general discussion of its place in the criminal law, see Glanville Williams, *Criminal Law, The General Part* (2nd edn.), Chap. II; and *The Mental Element in Crime,* Jerusalem, 1965), Chap. I.

Page 117. *Intention and recklessness.* For divergence of juristic usage contrast Bentham (loc. cit.) and Austin *Lectures on Jurisprudence* (5th edn.), p.424 with Glanville Williams (loc. cit.).

Page 118. *Bare intention.* For the importance of an intention to do a future act in the civil law, see Landlord and Tenant Act, 1927, s. 18 (1), and Landlord and Tenant Act, 1954, s. 30 (1), together with the discussion of the general nature of intention in *Cunliffe* v. *Goodman,* (1950), 2 Q.B. 237, at p. 253, and *Betty's Cafes Ltd.* v. *Phillips Furnishing Stores Ltd.,* (1957), Ch. 67.

Page 120. *Bentham on Direct and Oblique Intention:* see *Principles of Morals and Legislation.* Ch. 8, para vi, 'A consequence when it is intentional may be either *directly* so or only *obliquely*. It may be said to be directly or linearly intentional, when the prospect of producing it constituted one of the links in the chains of causes by which the person was determined to do the act. It may be said to be obliquely or collaterally intentional when although the consequence was in contemplation, and appeared likely to ensue in case of the acts being performed, yet the prospect of producing such consequences did not form a link in the aforesaid chain.' For the same distinction in Continental legal codes, see e.g. Italian Penal Code, Art. 43, *delitto doloso,* contrasted with *delitto colposo.*

English theoretical writers on the criminal law draw virtually the same distinction in a different terminology by defining intention for legal purposes in terms either of desire for certain consequences or 'foresight of certainty': see Glanville Williams, loc. cit. But the concepts of desire and intention need to be distinguished for many different purposes even in the law (see *Lang* v. *Lang*, (1955), A.C. 402, cf. *Gollins* v. *Gollins*, (1964), A.C. 644), and unless 'desire' is given a technically restricted meaning, the analysis of intention in terms of desire may be misleading.

Page 122. *The Doctrine of Double effect in Catholic moral theology*. See, for a searching examination of an important aspect of this doctrine in relation to the termination of pregnancy, Jonathan Bennett, 'Whatever the Consequences', *Analysis*, Vol. 26 (1966), p. 83, and cf. Glanville Williams, *The Sanctity of Human Life and the Criminal Law*, p. 177, *et seq*. For details of the doctrine in its application to medical problems, see J. B. McAllister, *Ethics with Special Application to the Medical and Nursing Professions*, and G. Kelly, S.J., *Medico-Moral Problems* (Dublin, 1955). For its application to the relief of pain see 'The Papal Address to Doctors', (Feb. 24, 1957), *Catholic Medical Quarterly*, Vol. 10 (1957), p. 51.

Page 124. *Direct intention in abortion*. Killing the foetus by alteration of the amniotic fluid is a well-known technique. See letter to *The Times* Monday 5 Dec. 1966 from Mr. Edward Cope.

Page 125. *R*. v. *Steane*. For criticism of the interpretation given in this case to the words 'with the intention of assisting the enemy', see Glanville Williams, *Criminal Law, The General Part* (2nd edn.), p. 40 and *The Mental Element in Crime*, pp. 21–23.

Page 126. *Attempts and further intent*. For criticism of the analysis of the notion of an attempt in terms of direct intention, see Glanville Williams, *The Mental Element in Crime*, p. 24.

Page 127. *Bare intention*. For discussion of the reasons for requiring an overt act as a condition of liability for criminal punishment, see H. Morris, 'Punishment for Thoughts', *The Monist*, Vol. 49 (1965), p. 342.

Page 128. *The punishment of attempts on deterrent principles*. See Wechsler and Michael, 'A Rationale of the Law of Homicide', 37 *C.L.R.* (1937), at p. 1295. and Wechsler, Jones and Korn, 'The Treatment of Inchoate Crimes in the Model Penal Code of the A.L.I.', *C.L.R.* (1961), p. 573.

Page 129. *Less severe punishment for attempts*. In English law, though the maximum punishment for attempts laid down by statute is almost invariably less than for a completed crime (Sexual Offences Act, 1956, Sched. II Part I), an attempt, as a common law misdemeanour, is normally punishable with imprisonment at the discretion of the court, and although a sentence greater than the maximum allowed for the offence ought not to be imposed (Criminal Law Act, 1967, s. 7(2)) no judicial statement has been made *requiring* a lesser penalty than for the main offence. In most Continental systems and in many jurisdictions in the United States the maximum punishment for an attempt is less than that for the main offence. In the case of treason in English law, however, because of the exceedingly wide definition of it, it is doubtful whether a distinction can be drawn between an attempt and the completed crime. For conspiracy see *Verrier* v. *D.P.P.* (1967), 2 A.C. 195.

Page 130. *Beccaria on the punishment of attempts.* See *On Crimes and Punishments,* s.14.

Page 130. *The gradation of punishment by reference to the harm done.* For a defence of this principle on the ground 'that it gratifies a natural public feeling to choose out for punishment the one who has actually caused great harm', see Stephen, *History of the Criminal Law,* Vol. III, pp. 311, 312.

Page 131. *The severity of punishment and resentment.* Cf. with Bishop Butler's Sermon on Resentment, P. F. Strawson, 'Freedom and Resentment', *Proc. British Academy,* Vol. 48 (1962), p. 187. Durkheim, *The Division of Labour in Society,* Chap. II.

Page 133. *The punishment of negligence.* For discussion of sceptical doubts as to the efficacy of such punishment, see Chap. VI, *supra.*

Page 133. *Punishment and the deliberation of the potential criminal.* For the importance attached by the Utilitarians to the notion that the threat of punishment should operate in the calculations of the potential offender as a guiding reason, see *Seventh Report of the Criminal Law Commissioners,* 1843.

Page 135. *The House of Lords debate on causing death by dangerous driving.* See 191 *H.L.Deb.* 82–94 (1954–5), esp. col. 92.

CHAPTER VI

Page 137. *Negligence as the failure to take reasonable precautions against harm.* This commonly accepted general definition of what lawyers mean by negligence presents two features which frequently perplex the layman:

(i) In ordinary non-legal usage a deliberate or intentional failure to take reasonable precautions against harm would not be described as carelessness or negligence; for these expressions, as ordinarily used, carry the implication that the necessity for precautions was not appreciated and the failure to take them arose from failure to attend to and appreciate the risks. Lawyers, however, distinguish '*inadvertent* negligence' as one species of the more general notion, because there are certain legal contexts where negligence may be said to be deliberate or intentional, or at least accompanied by a clear appreciation of the risks involved. Thus, in the civil law, the fact that a defendant charged with negligently causing harm intended the harm, or appreciated the risks involved, is irrelevant to a finding of negligence (though it may in some cases increase the damages he will be ordered to pay), and the defendant in these circumstances may be held to have committed the tort of negligence even if he intended it. (Suggestions to the contrary made by Lord Denning in *Letang* v. *Cooper* (1964), 2 All. E.R. 1929 seem not well founded; see now *Long* v. *Hepworth* (1968), 3 All E.R. 248.) In the criminal law, however, where, as in the case of homicide, proof that the accused caused death negligently, rather than intentionally, has an extenuating effect, leading to a conviction for manslaughter rather than murder, negligence is understood to be 'inadvertent' and taken to exclude intention and recklessness in the sense of an appreciation of the gravity of the risks. But the offence of driving without due care or attention (Road Traffic Act 1960, s.2.) may apparently be committed deliberately, e.g. if a driver though appreciating the risks, averts his eyes from the road in crowded traffic in order to light a cigarette.

(ii) It should be observed that the expression 'failure to take precautions, though most frequently used in explicating the meaning of negligence in the law, would not be used ordinarily to refer to every type of negligence. This phrase is most appropriate to cover those cases where negligence consists in the careless *manner* in which the activity described by the verb is executed, e.g. driving a care carelessly, as when the driver fails to look at the road. 'Failure to take *precautions*' is, however, not naturally used, except by lawyers, where the negligence consists in doing something which, as a type of activity, is generally likely to cause harm, e.g. pointing a loaded gun at another person, which in non-legal use would hardly be described as failure to take precautions, though the extension of this expression to such cases occasions no difficulty in legal contexts once it is understood. The distinction between care or carelessness shown *in the manner* of executing an activity, and care or carelessness shown *by doing* certain things, is in English conveyed by the place of the adverbs 'carefully' or 'carelessly' before or after the verb: contrast 'He did X carelessly' with 'He carelessly did X'. For an illuminating discussion of various aspects of the concept of care and divergences between legal and non-legal uses of the associated expressions, see A. R. White, *Attention*, Ch. V, and the same author's controversy with P. J. Fitzgerald and Glanville Williams in *M.L.R.* (1961), p. 592 and in *M.L.R.* (1962), pp. 49, 55, 437. The position is, however, complicated, or at least obscured, by the fact that while some legal theorists confine the expression 'recklessness' to cases where the gravity of the risk is consciously appreciated, and treat reckless killing thus defined as murder (see Glanville Williams, *The Mental Element in Crime*, pp. 84–90, who criticizes law in this respect), others (e.g. Cross and Jones, *An Introduction to the Criminal Law*, 5th edn.) use the expression 'recklessness' to connote either gross inadvertent negligence or consciousness of the existence of a risk without appreciation of its gravity, and treat reckless killing as manslaughter. In the case of certain statutory crimes recklessness has been treated by the courts as equivalent to gross inadvertent negligence (see *R.* v. *Bates* (1955), 2 All.E.R. 842, and the contrary view taken in *R.* v. *Mackinnon* (1959), Q.B. 150.

Page 140. *Mens rea and capacities for control.* See the observation of Parker, C. J., in *R.* v. *Byrne*, *supra*, p. 192, and Notes, p. 241–2.

Page 141. *Voluntary conduct.* For detailed examination of this notion see Chap. IV, *supra.*

Page 148. *Degrees of negligence.* The criteria suggested here for the assessment of degrees of negligence are not exhaustive, and perhaps are only easily applicable where negligence consists in the failure to take what would be normally considered 'precautions' in the course of executing some activity not normally harmful. (See Notes, p. 260.) But negligence may well be considered 'gross' quite independently of the simplicity of the omitted precautions. This would be so where it consists in doing something which is obviously likely to cause harm in most circumstances, e.g. dynamiting a building in an inhabited area, even if the precautions required were onerous and elaborate or, indeed, if nothing that could ordinarily be called 'precautions' could make it safe.

Page 150. *Knowledge as a necessary and sufficient condition of the capacity for self-control.* See Hall, *Principles of Criminal Law,* and the references to the 'integration' theories of the mind, *supra*, p. 32.

Page 153. *Negligence and the individualization of liability.* The attempt to individualize the conditions of liability for negligence, so as to cater for those of less than normal capacities, physical or mental, presents more problems than are considered in the text. In some cases the two-stage test suggested on p. 154 would give unsatisfactory results by exempting from liability those who, while unable because of some personal disability to take the same precautions against harm as a normal 'reasonable man', yet could, and would, if reasonable, have taken some other precaution to avoid the same harm. This is so because certain incapacities or abnormalities can be intelligibly treated as factors or parts of the circumstances which a reasonable man would take into account in determining what was demanded by way of care. Thus, if a blind man of normal mentality walks out of his house into a busy road and knocks over a child passing on a bicycle at that moment, this might well be thought grossly negligent on his part; for though he could not have taken the same precautions as the ordinary sighted man (e.g. looked, seen, and waited), he could, and if thoughtful, would have, avoided the harm in other ways (e.g. by asking to be conducted across the road). But the two-stage test suggested at p. 154 would exempt all those who could not take the precautions which a sighted man would have taken.

So far as physical incapacities are concerned, the difficulty might be met by deleting from the first part of the two-stage test on p. 154, the words 'with normal capacities' and treating the expression 'in the circumstances' as including such physical incapacities which can be intelligibly regarded as factors with which a reasonable man would reckon. Findings of negligence may be intelligibly made on this basis because in relation to such physical incapacities there may either be stocks of common knowledge concerning the ways in which persons suffering from such disabilities do and can behave, or a judge or jury might, by imaginatively placing themselves in his position, intelligibly speculate as to the way in which a reasonable man, so afflicted would behave. Such stocks of common knowledge or imaginative speculations lie in the background of all adjudications upon negligence. But mental and psychological disabilities cannot be dealt with in this way for one or both of the following reasons.

(*i*) Very severe mental abnormality, or even gross stupidity, cannot without absurdity be treated as factors with which the reasonable man would reckon; for they are inconsistent with the minimum meaning of the supposition that he is reasonable, even though it is true that 'reasonableness' for this purpose is not purely a matter of intelligence.

(*ii*) Though lesser mental abnormalities might be attributed without absurdity to the reasonable man, judgements as to the way he would, in spite of his afflictions, have behaved in order to avoid harm will, in most cases, be impossible for others to make, at least until medical science has built up some stocks of knowledge on the subject, comparable to those which guide judgements on negligence in ordinary cases. In the case of mental disability a judge or jury's speculation as to how they would have behaved themselves, if similarly afflicted, would in most cases be worthless.

The choice therefore would seem to lie between (*a*) exempting from criminal liability for negligence all those whose mental disabilities were such to prevent them taking the precautions that the ordinary man would have taken, thus foregoing any speculation as to whether they could, in spite of their affliction have taken some different but adequate precaution, or (*b*) exempting from criminal liability for negligence all persons suffering from specified types of mental abnormality. In practice the operation of the Mental Health Act, 1959, s. 60, will usually lead to the latter result. It should be observed, however, that no case is known where a person accused of manslaughter by criminal negligence, or any other form of criminal negligence, has pleaded insanity, and it is not clear how the M'Naghten Rules would apply to such a case (see the discussion of this by Glanville Williams, *Criminal Law, The General Part* (2nd edn.), p. 101 and pp. 527–9.

The individualization of a standard of care in civil negligence has been more exhaustively discussed. See, for an illuminating discussion of this problem and an examination of the force of the distinctions between 'subjective' and 'objective' in relation to it, Seavey: 'Negligence—Subjective or Objective', 41 *Harvard Law Review* (1927), pp. 1–28.

Page 156. For the character of sceptical doubts concerning the efficacy of punishment for negligence see *supra*, p. 133-4.

CHAPTER VII

Page 159. *Inconsistent ideas in sentencing.* For further discussion of the principles of sentencing see *The Report of the Interdepartmental Committee on the Business of the Criminal Courts* (The Streatfield Report, H.M.S.O., 1961, Cmd. 1289), and the observations thereon in Wootton, *Crime and the Criminal Law*, Chap. IV. On the recommendation of the Streatfield Committee the Home Office prepared *The Sentence of the Court* (2nd edn., H.M.S.O., 1969), a handbook to help courts in the selection of sentences. See also Walker, *Sentencing in a Rational Society* (1969).

Page 161. *Bentham and the proportionality of punishment.* Bentham criticized Beccaria for failing to subject the idea of proportion to any critical analysis, and developed his own utilitarian interpretation of the idea in *Principles of Penal Law,* Pt. 2, Bk. 1, Ch. VI (*Works,* Vol. 1., pp. 399–402) and *Introduction to the Principles of Morals and Legislation,* Ch. XVI.

Page 162. *Harm done as the measure of seriousness of crime.* See for criticism of this idea, Whiteley, 'On Retribution', *Philosophy* (1956), p. 154 and *supra,* pp. 129–31, 155, and Notes, p. 259.

Page 166. *The 'Double-Track' system and the distinction between punishment and social protection.* For a recent defence of this distinction, see Sparks, 'Custodial Training Sentences: Another view', *Crim.L.R.* (1966), p. 8.

Page 166. *Preventive detention.* It is generally considered that the system of preventive detention provided by the Prevention of Crime Act, 1908, and the Criminal Justice Act, 1948, has not been successful, since it has resulted in the lengthy and costly incarceration of relatively feeble offenders persisting in petty crime, rather than those who constitute a real menace to society. A new scheme is adumbrated in the Government White Paper, *The Adult Offender* (H.M.S.O., 1965, Cmd. 2852), paras. 9–17. But see the criticism by Cross in 'Penal Reform in 1965', *Crim.L.R.* (1966), pp. 191–4.

Page 168. *Judges as sentencers.* For various criticisms of the present system and suggestions for change see Walker, 'The Sentence of the Court', in *The Listener,* 28 June 1962, Cross 'Indeterminate Sentences', in *The Listener,* 15 Feb. 1962, and Wootton, *Crime and the Criminal Law,* Ch. IV.

Page 170. *The denunciatory theory of punishment.* In Lord Denning's version of this theory, 'the emphatic denunciation by the community of a crime' (as contrasted with deterrence, prevention, or reform) is identified as the *ultimate justification* of punishment. Accordingly, in spite of many similarities, this version of the theory needs to be distinguished from those which treat the expression of the community's condemnation not as a justification but as a defining feature of legal punishment. Thus Feinberg 'The Expressive Function of Punishment', *The Monist* (1966), p. 307 defines punishment as a conventional device for the expression of attitudes of resentment and indignation, and then proceeds to consider the justification of punishment in terms of certain desirable social consequences which the expression of condemnation and indignation are thought to have.

Page 178. *More or less extreme versions of the proposal to eliminate responsibility.* Lady Wootton in her latest book (*Crime and the Criminal Law*) explicitly advocates the more extreme version in which all reference to a mental element would be eliminated from the definition of offences. See, for a detailed account and criticism of this version, Chap. VIII.

Page 181. *Ignorance of the law.* Not all legal systems insist on the rule *ignorantia legis neminem excusat* as severely as Anglo-American law, and the argument urged by Hall that the rule is a necessary consequence of the character of legal systems (*General Principles of Criminal Law* (2nd edn.) pp. 382–3) seems mistaken. See J. Andenaes, '*Error Juris* in Scandinavian Criminal Law', in *Essays in Criminal Science,* ed, G. Mueller.

Page 184. *Criminal responsibility in children.* The age of responsibility was raised to 10 by the Children's and Young Persons Act, 1963. See now Children and Young Persons Act, 1969, excluding criminal proceedings, except for homicide, against children under 14.

Page 184. *Mental Health Act,* 1959, *s.* 60. See for further discussion of the operation of this section, *supra,* p. 198 and Notes, p. 264.

CHAPTER VIII

Page 188. *Holmes and the doctrine of objective liability.* See *supra,* p. 38 and Notes, p. 242–4.

Page 196. *The meaning of responsibility.* The distinction drawn in the text between two meanings of legal responsibility, respectively designated as 'legal accountability' and 'personal responsibility', has been challenged by Simpson in a review of this lecture, (see *Crim.L.R.* (1966), p. 124). For a consideration of this and other criticisms and a more comprehensive account of the notion of responsibility see *supra,* pp. 212-30.

Page 198. *The Mental Health Act 1959, s. 60.* For an account of the operation of the provision for the mentally abnormal made by this act see McCabe, Rollins, and Walker, 'The Offender and the Mental Health Act', in *Medicine, Science and Law* (1964), p. 231, and Walker, 'The Mentally Abnormal Offender in the English Penal System', *Soc. Review Monograph,* No. 9 (June 1965).

Page 204. *Circular character of the inference of impaired capacity from conduct.* For criticisms of the contention that such inferences are necessarily circular, see V. Haksar, 'The Responsibility of Psychopaths', *Philosophical Quarterly* (1965), p. 135; F. Jacobs, 'Circularity and Responsibility', *Philosophy* (1964), p. 268; and Walker 'Psychopathy in Law and Logic' in *Medicine, Science and Law* (1965), p. 3.

Page 209. *Mens rea in the definition of attempts.* Lady Wootton's proposals to eliminate all reference to a mental element in the definition of offences would presumably involve substituting, for the notion of an act *intended* to have a certain consequence, which is involved in the concept of an attempt, the notion of an act *likely* to have such a consequence. The courts, however, at present seem reluctant to take such a step (see *Gardner* v. *Akeroyd* (1952), 2 Q.B. 743 where the meaning of the expression 'an act preparatory to committing an offence' was in issue.

CHAPTER IX

Page 212. *Different senses of responsibility.* I have not considered in the text whether there is any unifying feature which explains the diverse applications of the word 'responsibility'. Etymology suggests that the notion of an 'answer' may play the part: a person who is responsible for something may be required to answer questions and it has been often pointed out that traces of this survive in some of the senses of responsibility. To say that a minister is responsible for the conduct of his department implies that he is obliged to answer questions about it if things go wrong, and perhaps in all cases of role-responsibility there is an obligation to answer such questions. But in the case of causal responsibility the notion of answering questions seems not to be involved. I think, therefore, that though some sense of 'answer' is connected with all the main meanings of responsibility it is not that of answering questions, and the connexion though systematic is indirect.

The following account seems to me to be plausible. The original meaning of the word 'answer', like that of the Greek "αποκρινέσθαι' and the Latin *respondere,* was not that of answering questions, but that of answering or rebutting accusations or charges, which, if established, carried liability to punishment or blame or other adverse treatment (see *O.E.D., sub. tit.* 'answer'). There is, therefore, a very direct connexion between the notion of answering in this sense and liability-responsibility, which I take to be the primary sense of responsibility: a person who fails to rebut a charge is liable to punishment or blame for what he has done, and a person who is liable to punishment or blame has had a charge to rebut and failed to rebut it. Hence it was once common in legal and other usage to speak of a person as answerable for loss or damage (see Pollock, *Torts* (3rd edn., 1892), pp. 432–44, 465), and also as answerable for his actions, in cases where we should use the expression 'responsible for'.

The other senses of responsibility are variously derived from this primary sense of liability-responsibility and are connected indirectly with the relevant sense of answer in that way. Causing harm and the possession of the normal capacities to conform to the requirements of law or morals are the most prominent among the criteria of liability-responsibility. So a person who causes harm by his action or omission, and possesses these capacities, is responsible in the liability-responsibility sense. It seems, then, altogether natural that the word 'responsible' should be used not only for this result of satisfying the criteria but also of what satisfies them (a person causing harm, a person having the normal capacities). It seems a further, quite natural extension that 'responsible for' should be used to signify causal connexion and the possession of these capacities outside contexts of blame or punishment, and, in the case of causal responsibility, that it should be extended to the production of good outcomes as well as bad, and to other causes besides human beings and their actions. Role-responsibility is perhaps less directly derivable from the primary sense of liability-responsibility: the connexion is that the occupant of a role is contingently responsible in that primary sense if he fails to fulfil the duties which define his role and which are hence his responsibilities.

Page 216. *Legal writers' usage of 'responsibility' and 'liability'*. 'Strict responsibility' and 'vicarious responsibility' are the titles of Chapters VI and VII of Glanville Williams, *The Criminal Law, The General Part,* 2nd edn. Cf. Morris and Howard, *Studies in the Criminal Law,* Chap. VI, entitled 'Strict Responsibility' and the last-mentioned author's book of the same title. This use of 'strict responsibility' is no modernism, see Pollock on *Torts* (3rd edn. 1892), p. 434.

Smith and Hogan in their *Criminal Law* use 'strict liability' and 'vicarious liability' in titles for their Chapters VI and VII, s. 2, but refer frequently in the text to 'vicarious responsibility' (e.g. pp. 87–89). Cf. Bayer, 'Criminal Responsibility for the Acts of Others', *43 Harvard L.R.* (1930), p. 689. and Prosser, *Torts,* (2nd edn.) s. 62: 'Vicarious Liability is the responsibility of one person, without any wrongful conduct of his own for the acts of another'.

Page 219. *Distinction between capacity for the control of conduct and other elements of mens rea.* The German Penal Code, Art. 51 (as amended in 1933), exempts under the heading 'Zuschreibungunfähigkeit' (meaning, literally, incapacity for ascription), those who lack the ability to recognize the wrongness of conduct and to act in accordance with that recognition, and distinguishes this from lack of particular elements of knowledge or other subjective elements required by the definition of particular offences. The Italian Penal Code, Art. 85, declares that only persons who possess the capacities of intending and willing (*intendere* and *volere*) are *imputabile* (generally translated 'responsible') and distinguishes the lack of knowledge or intention on the part of a normal person on particular occasions as matters relating to *dolo* (Arts. 42 and 43), though the phrase *responsibilita per dolo* is used in the rubric to Article 42. Bentham, in Chapter XIII of *The Principles of Morals and Legislation,* in discussing cases 'not mete for punishment', distinguishes between those where the law's threats could not be effective because of the agent's standing incapacity, due to more or less persisting conditions such as insanity, infancy, or intoxication, and cases where the law's threats are ineffective because the agent who is endowed with the normal capacities is under some mistake, or acts unintentionally, or is subjected to duress on a particular occasion. Any such distinction will occasionally produce cases difficult to classify: thus subjection to duress or sudden lapse of consciousness (automatism) may be regarded either as a short-lived incapacity of the agent or as a defect on a particular occasion, of his action.

Page 219. *Extension of the expression mens rea to normal capacities of control.* See Stroud, *Mens Rea* (1914), which includes a chapter on insanity; Cross and Jones, *Introduction to Criminal Law,* 5th edn., pp. 31–34, which refers to insanity as part of the topic of *mens rea.* For a similar American usage see Hall, *Principles of Criminal Law,* and for judicial usage see *U.S.* v. *Currens,* 290, F. 2nd. 751 (Biggs, C. J.). Perhaps the dominant trend of English legal writing treats insanity as a topic distinct from *mens rea.*

Page 231. *Varieties of retributive theory.* For a modern review of these see Walker, *Sentencing in a Rational Society* (1969).

Page 233. *Proportion.* For Bentham's treatment of proportion see Notes, p. 263.
Gravity of harm done. For discussion of this criterion of severity of punishment, see *supra,* p. 130.

Page 235. *Punishment as the reinforcement of social morality.* See Durkheim, *The Division of Labour in Society,* Chap. II, and the remarkably similar view of the English judiciary as represented by Lord Denning (*supra,* p. 170) and Lord Devlin, *The Enforcement of Morals* (1965).

Page 236. *Division of the field between retributive and utilitarian theories.* See Lord Devlin's essay, 'Morals and the Quasi-criminal Law and the Law of Tort' in *The Enforcement of Morals,* for the argument that different principles should be applied to crimes which overlap with moral offences and to those (quasi-crimes) which do not. For criticism of this view, see Fitzgerald, 'Crimes and Quasi-Crimes', 10 *Natural Law Forum* (1963).

Page 237. *Retribution as setting a maximum for punishments.* For this relaxed form of retributive theory see Longford, *The Idea of Punishment* (1961).

APPENDIX

Continuance of the Murder Act, 1965, and Murder Statistics 1957–68, England and U.S.A.

1. A resolution to the effect that the Murder Act, 1965, should not expire was passed by the House of Commons on 16 December 1969 by 343 votes to 185; and a similar resolution was passed by the House of Lords on 18 December 1969 by 220 votes to 174. Advocates and opponents of the death penalty offered, in and out of Parliament, conflicting interpretations of the statistics available since 1965 as indications of the effectiveness of the death penalty. The principal 'statistical' arguments in support of the retention of the death penalty were based on the figures for 'capital' murder (estimated and actual) for the period 1957–68, and on the figures for various crimes of violence other than murder, and for crimes in which firearms were used. (For the debates on the resolutions see 793 *H.C. Deb.* 939 ff. and 1148 ff. and 306 *H.L. Deb.* 1106 ff.) This note, which is mainly based on Gibson and Klein, *Murder 1957 to 1968*, a Home Office Statistical Division Report (H.M.S.O., 1969), is designed to summarize the principal statistical facts bearing on this controversy.

2. The Murder Act, 1965, came into force on 9 November 1965. Five persons were sentenced to death in that year but all of them were reprieved before the Act came into force. 1965 was the first year in which there were no executions; there were two executions in 1964, although abolition was then under discussion. A second reading was given on 21 December 1964 by the House of Commons to the Bill which ultimately became law as the Murder Act, 1965.

3. Three principal sets of figures require attention, viz.:

(i) *Numbers and rates per million of population of murder and s. 2 manslaughter known to the police 1957–68.*

	No. of victims			*No. per million of home population of England and Wales*	
	Murder	*s. 2 Manslaughter*	*Total*	*Murder*	*Murder and s. 2 manslaughter*
1957	135	22	157	3.0	3.5
1958	114	29	143	2.5	3.2
1959	135	21	156	3.0	3.4
1960	123	31	154	2.7	3.4
1961	118	30	148	2.6	3.2
1962	129	42	171	2.8	3.7
1963	122	56	178	2.6	3.8
1964	135	35	170	2.8	3.6
1965	135	50	185	2.8	3.9
1966	122	65	187	2.5	3.9
1967	154	57	211	3.2	4.4
1968	148	57	205	3.0	4.2

(ii) *Numbers and percentages of the total of normal and abnormal murder and s. 2 manslaughter.* Murder classified as 'abnormal' includes (*a*) suspects found insane, (*b*) those convicted of s. 2 manslaughter, and (*c*) suspects who committed suicide.

	Abnormal homicide		*Normal murder*		*Total murder and s. 2 manslaughter*
	No.	%	*No.*	%	*No.*
1957	100	63.7	57	36.3	157
1958	96	67.2	47	32.8	143
1959	99	63.5	57	36.5	156
1960	103	66.9	51	33.1	154
1961	94	63.5	54	36.5	148
1962	115	67.3	56	32.7	171
1963	119	66.9	59	33.1	178
1964	94	55.3	76	44.7	170
1965	108	58.4	77	41.6	185
1966	99	53.0	88	47.0	187
1967	121	57.3	90	42.7	211
1968	109	53.2	96	46.8	205

(iii) *Estimated number of capital and non-capital murders.*[1] It is important to bear in mind Gibson and Klein's warning (op. cit., para. 20) that only those cases where the accused was actually convicted of capital murder can be certainly regarded as such. Since 1965 the classification of murder as 'capital' represents hypothetical speculation, on the basis of known circumstances, as to what would have been found if capital murder had been an issue at a trial which might have thrown doubt on those circumstances. (For an illustration of the probable exaggeration in such estimates, see Sir Edward Boyle in 793 *H.C. Deb.* at 1248–9.) Similar caution is needed as to the classification as 'capital' of murders before 1965 where the suspect was acquitted or found insane, or committed suicide, or the murder was unsolved. Special caution is needed in comparing the hypothetical figures for normal capital murder after 1965 with figures before 1965 where the classification as 'capital' was mainly made by a jury (Gibson and Klein, op. cit. p. 58, para. 76).

	Capital		*Non-capital*		*Total*
	No.	*% of total murders*	*No.*	*% of total murders*	*No.*
1957	25 (12)	18.5	110 (45)	81.5	135
1958	21 (11)	18.4	93 (36)	81.6	114
1959	23 (9)	17.0	112 (48)	83.0	135
1960	26 (11)	21.1	97 (40)	78.9	123
1961	18 (7)	15.3	100 (47)	84.7	118
1962	18 (4)	14.0	111 (52)	86.0	129
1963	14 (7)	11.5	108 (52)	88.5	122
1964	17 (10)	12.6	118 (66)	87.4	135
1965	30 (17)	22.2	105 (60)	77.8	135
1966	36 (29)	29.5	86 (59)	70.5	122
1967	46 (24)	29.9	108 (66)	70.1	154
1968	42 (26)	28.4	106 (70)	71.6	148

(Figures in brackets represent numbers of normal murders.)

[1] 'Murder' in this section does not include s. 2 manslaughter.

4. The salient points which emerge from the study of the statistics are as follows:

(i) There has been since 1962 a rise in the combined numbers and rates of murder and s. 2 manslaughter. From 1957 to 1961 the number of these offences and the rates per million of population varied between 143 and 157 (3.2 and 3.5), but from 1962 to 1968 they varied between 170 and 211 (3.6 and 4.4). The largest annual increase in numbers during 1957–68 was in 1967: an increase of 24 over the previous year's figure of 187; the increase in rate per million (0.5) over the previous year's rate was the largest since 1962 (also 0.5).

(ii) Since 1964 normal murder has risen sharply both in numbers and as a percentage of the total of murders and s. 2 manslaughter. From 1957 to 1963 the numbers of normal murder varied between 47 and 59, and percentages of the total between 32.7 per cent and 36.5 per cent; and from 1964 to 1968 numbers varied between 76 and 96, and percentages of the total between 41.6 per cent and 47 per cent. No reason is known for this increase. The largest annual increase was in 1964, both in numbers and percentages of the total of murder and s. 2 manslaughter: an increase of 17 over the previous year's figure of 59, and of 11.6 per cent over the previous year's percentage of the total (33.1 per cent).

(iii) (*a*) From 1957 to 1964 capital murders (actual and estimated figures) varied between 14 and 26 (11.5–21.1 per cent of all murders) and from 1965 to 1968 (estimated figures) between 30 and 46 (22.2–29.9 per cent). The largest annual increase during the years 1957–68 both in numbers and percentages of the total was in 1965, an increase of 13 over the previous year's figure of 17 and of 9.6 per cent over the previous year's percentage of the total (12.6 per cent).

(*b*) In the four-year period 1965–8, the estimated figures for capital murder as a whole and also for normal capital murder were considerably larger than the figures (actual and estimated) for the two previous periods of four years. The numbers of non-capital murders as a whole for the period 1965–8 were less than for the two previous periods of four years, and although the figures for normal non-capital murder for 1965–8 were larger than for the two previous periods of four years the percentage increase was far less than in the case of normal capital murder. The relevant figures are as follows:

	1957–60			*1961–4*			*1965–8*		
I. Capital murder	95	normal	43	67	normal	28	154	normal	96
		abnormal	52		abnormal	39		abnormal	58
II. Non-capital murder	412	normal	169	437	normal	217	405	normal	255
		abnormal	243		abnormal	220		abnormal	150

However, comparable figures for Scotland (see Gibson and Klein, op. cit., p. 84) show different and in some respects opposite trends. Thus figures for both capital and non-capital murder for the four-year period 1965–8 were greater than for the two previous four-year periods, but the percentage increase for 1965–8 over the period 1961–4 was approximately the same for both capital and non-capital murder. The percentage increase of normal capital murder for 1965–8 over the previous four-year period was far less than the percentage increase for normal non-capital murder. The relevant figures are as follows:

	1957–60			*1961–4*			*1965–8*		
I. Capital murder	19	normal	13	11	normal	8	20	normal	12
		abnormal	6		abnormal	3		abnormal	8
II. Non-capital murder	44	normal	24	66	normal	30	119	normal	98
		abnormal	20		abnormal	36		abnormal	21

(*c*) Gibson and Klein (op. cit., p. 13, para. 23, and p. 58, para. 73) consider that the rise in both normal capital and normal non-capital murder started in 1964. But in 1965 and 1966 the annual increase in numbers (estimated) of normal capital murder, and in their proportion of all normal murder, was greater than the increase in any previous year; whereas the number of non-capital normal murders, and their proportion of all normal murders, decreased in both these years. On the other hand, capital murder formed a larger proportion of abnormal than of normal murder in six of the twelve years 1957–68 (viz. 1959, 1961, 1962, 1965, 1967, and 1968). The largest single category of capital murder was murder by shooting followed by suicide (see 793 *H.C. Deb.* 1156).

5. *Trends in Indictable Crimes of Violence generally*

(*a*) Indictable crimes of violence against the person known to the police increased constantly between 1957, when it was 244 per million of population, and 1968, when it was 655 per million of population. (See Gibson and Klein, op. cit., p. 5, Table 3.) The percentage increase in each of the years 1965 to 1968 over the previous year was less than the percentage annual increase in any of the years 1957–64, except 1962. From 1955 there has been a constant and large increase in malicious woundings known to the police. The figure for 1968 represented an increase of 364 per cent over the figure for 1955. Malicious wounding comprises the less serious offences against the person and is to be distinguished from felonious wounding mentioned in (*b*) *infra*. (See McClintock and Avison, *Crime in England and Wales* (London, 1968), for these trends.)

(*b*) The following account of percentage increases are calculated from figures given in *Criminal Statistics for England and Wales 1968*, p. 2, Table A. The annual average figure for malicious wounding for the four years 1965–8 represents a 55 per cent increase over the average figure for the five years 1960–4; but this last figure represents an increase of 86 per cent over the average for the preceding five years, 1955–9. Similarly, though the average annual figure for wounding and other acts endangering life (felonious wounding) for 1965–8 represents a 24 per cent increase over the average figure for 1960–4, the latter figure represents a 37 per cent increase over the average figure for 1955–9. On the other hand, the annual average figure for attempted murder for 1965–8 represents a 24 per cent increase over the average for 1960–4, whereas the latter figure represents only a 19.5 per cent increase over the average for 1955–9.

(*c*) Statements in the Report of the Commissioner of Police of the Metropolis for 1968 on the use of firearms in indictable offences were quoted in debate on the resolutions for the continuance of the Murder Act, 1965 (793 *H.C. Deb.* 1178, 1210). According to these statements there was a rise in 1968 of 17 per cent over 1967 in indictable offences in which firearms[1] were used, and of 31 per cent in robberies or assaults with intent to rob where such weapons were used. The corresponding figures for 1967 were *decreases* of 8.9 per cent and of 9.3 per cent in comparison with 1966, and for 1966, increases of 14.2 per cent and 35.3 per cent over 1965. In 1965 the figures given for robberies in which firearms were known to have been carried were 114 as compared with 92 for 1964 (an increase of 24 per cent), and in 1964 were 92 as compared with 43 for 1963 (an increase of 114 per cent). (For these figures see the Reports of the Commissioner for the years in question.)

6. *Murder in the United States, 1960–8.* (All figures unless otherwise stated are taken from the Uniform Crime Reports.)

(*a*) The numbers and rates per million of population of murder and non-negligent manslaughter during these years were as follows:

	No. of offences	*Rate per million of population*
1960	9,000	50
1961	8,630	47
1962	8,430	45
1963	8,530	45
1964	9,250	48
1965	9,850	51
1966	10,920	56
1967	12,090	61
1968	13,650	65

The figures for 1968 represent an increase of 51.7 per cent in the number of offences and 36 per cent in rate over those for 1960. The corresponding increases in the numbers and rates of murder and s. 2 manslaughter in England and Wales were 30 and 20 per cent.

(*b*) Other comparisons between English and American figures (cf. p. 250 *supra*).

(i) Rates of state totals in Georgia and New Hampshire: in 1965–8 the rates in Georgia were 113, 113, 111, and 139 murders and non-negligent manslaughter per million of population; in New Hampshire the rates for the same years were 27, 19, 20, and 14.

[1] 'Firearms' includes supposed and imitation firearms. These were stated to be a small number in the Commissioner's Report of 1968 (p. 53).

(ii) Numbers and rates in Chicago: in 1965–8 the numbers of murders and non-negligent manslaughter were 455, 605, 647, and 732, representing rates per million of population of 69, 90, 95, and 107.

(iii) English figures: for the same years (1965–8) the numbers and rates per million of population of murders and s. 2 manslaughter in England and Wales were 185, 187, 211, and 205, representing rates per million of population of 3.9, 3.9, 4.4, and 4.2.

(*c*) Numbers of executions: in the ten years 1959–68 the number of executions in the United States for all crimes in these years were 41, 54, 33, 41, 18, 9, 7, 1, 2, and 0. (Figures given by the Lord Chancellor in the Lords debate on the Resolutions: 306 *H.L. Deb.* 1119.)

INDEX

(N.B. References to pages after p. 237 are to the Notes)